PRAISE FOR *WHAT YOUR BODY KNOWS ABOUT HAPPINESS*

"A fresh and engaging perspective on happiness from the inside out. Reading this book will make you happier, and living out its principles will help you thrive!"
—Aditi Nerurkar, MD, author of *The 5 Resets*

"This book brings research out of psychology and neuroscience labs and takes it for a walk in the real world, allowing us to overcome the disconnect between our mind and our body in the process. Delightful and well researched, *What Your Body Knows About Happiness* will help you find more ways to explore happiness in everyday life."
—Shawn Achor, *New York Times* bestselling author of *Big Potential* and *The Happiness Advantage*

"Grab this book, find a comfortable chair, and settle in for a delightful master class in what makes us happy. Grounded in science, Janice Kaplan's book will show you how the connections between mind and body—hidden in plain sight—can be harnessed every day to bring more joy into your life."
—Robert J. Waldinger, MD, *New York Times* bestselling co-author of *The Good Life* and Professor of Psychiatry, Harvard Medical School

"My grandmother always told me 'Healthy mind in a healthy body.' Janice Kaplan's terrific book is an exploration of why my grandma turned out to be right. The mind and body cannot be separated. Janice uses science, personal stories, and humor to show how we can live happier, more fulfilling lives by paying attention not only to our brains, but also everything below our necks. Everybody (and every mind) can benefit from reading it."
—AJ Jacobs, New York Times bestselling author of *The Year of Living Biblically* and *It's All Relative*

"*What Your Body Knows About Happiness* is an exploration of the importance of our bodies in experiencing true joy. Through engaging storytelling and scientific research, Janice Kaplan uncovers the wisdom encoded within our physical selves and offers practical strategies to harness this knowledge for a happier, more fulfilling life."
—Ellen Langer, Professor of Psychology, Harvard University and author of *The Mindful Body*

"A must-read for anyone looking for practical ways to enhance their daily joy. Kaplan's exploration beautifully explores the relationship between mind and body, showing how intertwined the two are in supporting our well-being. By blending engaging storytelling with the latest research, Kapan offers valuable insights into how making simple changes in either your routine, physical environment, or both, can have a profound positive impact on your happiness."

—Mike Rucker, PhD, author of *The Fun Habit*

"Our bodies influence our minds and our minds watch out for our bodies—after all, they grew up together! Janice Kaplan's delightful book brings to life all the wonderful, and often surprising, evidence of this close bond between two lifelong friends."

—John Bargh, author of *Before You Know It* and Professor of Psychology, Yale University

What Your Body Knows About Happiness

HOW TO Use Your Body TO Change Your Mind

JANICE KAPLAN

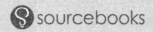

Copyright © 2025 by Janice Kaplan
Cover and internal design © 2025 by Sourcebooks
Cover Design by Jillian Rahn/Sourcebooks
Internal images © dstarky/Adobe Stock
Internal design by Tara Jaggers/Sourcebooks

Sourcebooks and the colophon are registered trademarks of Sourcebooks.

All rights reserved. No part of this book may be reproduced in any form or by any electronic or mechanical means including information storage and retrieval systems—except in the case of brief quotations embodied in critical articles or reviews—without permission in writing from its publisher, Sourcebooks.

This publication is designed to provide accurate and authoritative information in regard to the subject matter covered. It is sold with the understanding that the publisher is not engaged in rendering legal, accounting, or other professional service. If legal advice or other expert assistance is required, the services of a competent professional person should be sought.—*From a Declaration of Principles Jointly Adopted by a Committee of the American Bar Association and a Committee of Publishers and Associations*

References to internet websites (URLs) were accurate at the time of writing. Neither the author nor Sourcebooks is responsible for URLs that may have expired or changed since the manuscript was prepared.

This book is not intended as a substitute for medical advice from a qualified physician. The intent of this book is to provide accurate general information in regard to the subject matter covered. If medical advice or other expert help is needed, the services of an appropriate medical professional should be sought.

Published by Sourcebooks
P.O. Box 4410, Naperville, Illinois 60567-4410
(630) 961-3900
sourcebooks.com

Cataloging-in-Publication Data is on file with the Library of Congress.

Printed and bound in the United States of America.
MA 10 9 8 7 6 5 4 3 2 1

To Jacob, Eli, Ben, and Noor
beacons of happiness

Contents

INTRODUCTION   XI

Part One: The Science of Happy Bodies

   1: What Your Apple Watch Can't Tell You   1

   2: How Your Body Makes You Happy   20

   3: Your Mixed-Up Mind   39

   4: How Your Senses Give You Joy   60

Part Two: The Best Places on Earth to Be Happy

   5: Why Blue and Green Are the Happiest Colors   81

   6: Places That Make Your Spirits Soar   103

   7: Why Wine Tastes Better in Paris   120

Part Three: The Unexpected Power of Sex, Exercise, and Diet

   8: What Body Positivity Really Means   141

   9: The Happy Body Food Plan   167

   10: How Exercise Makes You Happy   193

Part Four: How Your Brain Resolves Pain

    11: Everybody Hurts (Sometimes) — 215

    12: Pain, Pain, Go Away — 234

    13: Sugar Pills Are Sweeter Than You Think — 250

Part Five: Why Pleasure and Creativity Come from Within

    14: The Neuroscience of Invention — 271

    15: How Your Body Makes You Smart — 288

Part Six: Your Optimistic Body

    16: What Language Does Your Body Talk? — 311

    17: The Body-Mind Happiness Plan — 328

SUGGESTED READING — 343

ACKNOWLEDGMENTS — 345

NOTES — 347

INDEX — 367

ABOUT THE AUTHOR — 377

Introduction

TRAVELING TO A VACATION RECENTLY, I faced an endless day of lost luggage and inexplicable airport delays. I stood in the tiny airport where my husband and I finally landed feeling frustrated and grumpy. In desperate need of a more positive view right then, I decided to try a little gratitude exercise.

I took a deep breath. I could handle this.

"Our luggage is lost, but at least I have a bathing suit in my carry-on bag!" I said brightly to my husband.

He laughed, and I calmed down enough to get out of the airport without bursting into tears. But it wasn't until we finally got onto a small motorboat to whisk us to our island location that I truly relaxed. The sun sparkled on the ocean and the warm air blew gently against my face. Barely ten minutes on the water and I felt blissfully happy—as if nothing frustrating at all had occurred. The sudden elation felt wonderful, and it also got me thinking about what was happening. In this setting, I didn't need to make a conscious decision to be grateful and positive. My body was sending the happiness message on its own.

Most discussions about happiness suggest that joy and pleasure exist only in our heads and that we have the power to control how we feel in any situation. I have been part of that conversation. In talks I give all over the country about my bestselling book *The Gratitude Diaries*, I explain that a key to being happy is changing your perspective—much as I did in the airport. I offer tips on how to reframe any situation or life event so you can see the good in it, and I'm glad that I have helped many people find gratitude in even tough times.

But at that moment on the boat, I understood how important it is to put ourselves in situations where our bodies can respond as positively as our minds. I was struck by the realization that happiness isn't exclusively a conscious decision. It comes from every fiber, sinew, and cell in our bodies. Very often our bodies send signals about how we feel and our conscious brains are simply *responding* rather than in control.

That recognition can be shocking. We usually think of our conscious minds as being the only source of our feelings and emotions. We are in full control of how we feel and what we do. But that view means we are missing out on the power of our bodies to be part of the conversation. Comic John Mulaney has joked that he doesn't know what his body is for other than taking his head from room to room. Mulaney gets a big laugh with the line because it's so on the nose with how most of us feel. We understand we're supposed to take care of our bodies in order to stay healthy, but beyond that, they seem slightly separate from the true person we are. Unless you're a professional athlete, you probably agree with Mulaney that your body is mostly a transport system for the computer sitting on top of your shoulders.

Most of us think of all the things that our bodies fail to do—they aren't strong enough, they get aches and pains, they fall prey to viruses and bacteria, they fail us as we get older. The various problems that occur too often put us at odds with our bodies. We see them as a challenge to overcome rather than a partner in making us happy. But the truth is that your body *wants* you to be happy and well. Its whole job is to send you the information that will allow you to thrive, and it never stops feeding your brain all the sensory input that it possibly can. Evolutionary biologists tell us that the primary goal for any species—humans, dogs, apes, ants, or crocodiles—is to be strong enough to reproduce and care for their young. In humans, that requires a large amount of time and energy, and your body works incredibly hard to keep itself in balance and peak form.

Once we understand what our bodies are doing to help us be happier, we can contribute to the cause rather than undermine it. Your body has a host of happiness hormones that get released in many situations from exercise to sex—and they make you calmer, less stressed, and more joyful. Ironically, some of the most dangerously addictive drugs in the world, including the opioids that have led to a devastating crisis of addiction in the last decade, simply imitate what our bodies can naturally provide. When these feel-good chemicals come in their natural form, they are helpful, not destructive.

This is not to say that our bodies always get it right. Antibiotics, vaccines, insulin, surgical procedures, and other medical breakthroughs have saved lives and transformed our longevity and very existence. Part of being human is the ability to change and improve ourselves and our world. But in doing that, we can't lose sight of the basic senses that inform us, the hormones that affect our moods, and the neurological systems that allow us to feel pain and pleasure.

We inhabit our bodies, but it's amazing how little we really understand about how they function. Neuroscientists are spending more and more time grappling with how the brain works, but many of them also recognize that our mind and body are in partnership. You can't have or understand one without the other.

The Body-First Approach

I used to envision the "me" that is "me" to be somewhere in my head. I first started to consider another view when I was writing *The Gratitude Diaries* and realized that going outside for a walk or enjoying a beautiful sunset could change how I felt—making me more grateful and calm. Since then I have thought more and more about how our bodies influence our happiness and well-being on a daily basis. The problem is that we don't always realize what is going on—and so we miss the chance to make our lives better and richer.

Often when we think we are making rational decisions,

Introduction

we are actually responding to signals from our body. Antonio Damasio, a professor of psychology, philosophy, and neurology at the University of Southern California, has written extensively about the sources of emotions and action. He says that we like to think of our brain as the decider, but it's really not. "Without our bodies there can be no consciousness," he says. The brain is simply acting as one of the two essential partners—body and brain—that formulate our behavior.

Our bodies send us information through hundreds of subtle clues like the release of hormones, a change in blood pressure, and the rise or fall of skin temperature. If your heart starts pounding when you turn onto a dark street at night, you'll probably pick a different path. But here's a question to think about: Does your heart start pounding because you're scared, or are you scared because your heart starts pounding? If you're a mind-centered person (like most of us are), the question seems almost silly. Of course your brain tells you you're scared and the physical response follows! But many scientists say we have made a mistake believing our brain is in charge—when it's really body first.

One of the most compelling theories about behavior says that you construct your emotions based on the biological processes going on within your brain and body. A whole new academic field is building up around interoception—your brain's interpretation of all the sensations going on in your body. Right now, for example, your heart is beating, your stomach is digesting, your lungs are filling with air, and your immune system is rushing out various hormones.

"Your body is part of your mind, not in some gauzy mystical way but in a very real biological way," says neurologist Lisa Feldman Barrett. "This means there is a piece of your body in every concept you make."

We all grew up being taught that you should keep a cool head in making decisions and not let emotions intrude. Now we are beginning to understand that approach is not only misguided, it's impossible. Emotions, feelings, and physical responses are part of everything we do. Attributing consciousness to the brain alone is a monstrous mistake, because, as Damasio says, "the brain is truly the body's captive audience," He calls it absurd to imagine having consciousness without a body.

French philosopher René Descartes brought to a wide audience in the seventeenth century the mistaken idea that mind and body are completely distinct. His much-repeated line "I think therefore I am" tells you which part of the mind-body duality he trusted. At the end of the nineteenth century, everything changed when William James, one of the most influential psychologists and philosophers of America (and brother of my favorite novelist, Henry James), turned Descartes upside down. He proposed a radically new theory of emotions, saying that *bodily changes lead to emotional feelings.*

James would have no question about what happens when you step onto that dark street at night. Your body reacts first with a pounding heart and surge of adrenaline. Your mind then scans the physical response and figures out—*I'm scared!* The emotion you experience is simply the conscious expression of what your body already knows.

This sounds great in theory, but it challenges everything we have come to understand about ourselves. We take pride in knowing that our brains are of a higher level than those of other animals—we're not exactly earthworms—but remember that the brain itself is a physical structure and part of the nervous system.

Let's think about that earthworm for a moment. It has sensory cells that send impulses to its brain and body, making sure it has proper responses to the environment. When it's too hot or cold, the earthworm burrows down into the ground to protect itself. When it senses that the conditions are okay, it extends its head up to the surface.

So yes, we're not earthworms, but we are still sensory creatures first. Like the earthworm, we have a central nervous system that responds to stimuli without any particular direction from the brain. If you accidentally touch a stove, you'll pull your hand away immediately, well before your brain has time to register what happened. When the doctor raps your knee with a hammer, your leg swings out. Knee-jerk response? You bet. What William James and Damasio and others want us to understand is that the instinctual response is just the first step. Your body's response *then tells your brain how to feel.* So for example, after the stove-touching, your conscious mind picks up that the central nervous system activated the motor neurons. Something happened that requires further consideration! Your brain takes the input and then fills in the emotion.

We all learned about the central nervous system in high school—but we never really think of our bodies acting without being directed by the brain. The brain seems to be the big computer telling us what to do and informing our bodies how to respond. But recognizing that the system can often work in reverse—our body programs the computer—is really incredibly exciting. It changes how you think about your everyday actions.

The brain contains many marvels, but it is not the be-all and end-all of who you are and what you do. The metaphor of brain as supercomputer doesn't hold, because the brain requires continuous reports from the body to function properly. A brain is not a computerized remote-control center that sends messages to a body, telling it how to behave. Your body and brain are intimately connected, sending signals back and forth every single moment of every single day. One body and one brain. *Together* they make you who you are.

Let's put it in some perspective. The brain weighs about 3 pounds, and if someone weighs (say) 150 pounds, that leaves about 147 pounds of them that is not brain. Nature doesn't waste prime real estate, whether on an animal, human, plant, or tree. The not-brain part of you is definitely around for some purpose. It happens all the time that you gasp or laugh or sneeze or cry and your brain rushes to catch up with the information that your body has provided. The body has the first response, and the conscious emotion will follow.

The Language of Happiness

As technology advances, we live increasingly in our minds and on our devices, our bodies an inconvenience to be cared for (when we remember) at gyms or with online videos. Instead of taking joy in our bodies, we grudgingly try to keep them healthy enough not to interfere with the rest of our lives.

We all know that we're supposed to listen to our bodies, but it's sometimes unclear what language they speak. Recent years have made the disconnection with our bodies even more dramatic. Working remotely from home rather than in an office and seeing fewer people in our daily interactions, we become ever more disembodied. Our voices drift over the internet on Zoom or Skype calls, unattached to any real physical being. We focus on our image from the shoulders up, the only part of us that can be seen in the computer's eye. On a call one day, a colleague mentioned that she had dressed up for our virtual meeting by wearing jeans. I admitted slight confusion at this new definition of dressing up.

"What do you usually wear?" I asked.

"Leggings or yoga pants or anything soft. Today, in your honor, I'm *structured*." We both laughed—and I had to admit that I spend many mornings at home writing in flannel pajama bottoms.

Clothes don't necessarily connect us to our physical selves in any positive way (though sometimes they can—a surprise we will discuss later), but my colleague picked the right word. Our bodies *do* provide the structure for our every mood and thought. As you join me in this book, you'll discover the cutting-edge research showing how mind and body can be a seamless team. Feelings like happiness, gratitude, and joy can emanate from our bodies as well as our minds.

If you have already learned how to make yourself happier through gratitude practices, you're a step ahead. Keep going! What you discover in this book will help you expand on those understandings. The new and exciting finding is that in addition to what our minds can achieve, we can identify bodily influences and environments that can inspire us to find our best and happiest selves.

Since each of us takes center stage in our own lives, it's important to understand that there are two stars in the show—body and mind. Please join me on the discovery of how we can use our physical selves and environs to change our creativity, outlook, success, and happiness. It's time to know how body and mind work together so that we can increase our joy every day—and find new ways to shine.

PART ONE

The Science of Happy Bodies

If you want to be happier, the place to start is in understanding your body. New research in fields from psychology to artificial intelligence reveals how powerfully our physical selves influence our moods and emotions.

What Your Apple Watch Can't Tell You

> Those who see any difference between
> soul and body have neither.
>
> —OSCAR WILDE

FOR A WHILE, EVERYBODY I knew was trying to walk ten thousand steps a day. It drove my husband crazy. He's a huge advocate of exercise and activity, but as a doctor, he also likes scientific data.

"You know there's absolutely no medical evidence behind that number," he grumbled.

He was right. The only scientific truth about ten thousand is that it's a nice (and memorable) round number. But his scientific approach was no match for the marketing skills of the Fitbit company, founded in 2007 by the determined entrepreneur James Park. After dropping out of Harvard (tech-boy cred!), Park began developing activity trackers you could strap to your wrist in pursuit of the mythic goal. By the time he sold Fitbit to Google in 2021 for more than $2 billion, tens of millions of Americans had bought both his product and his idea.

Park didn't originate the concept of ten thousand steps—a Japanese company that mass-produced a pedometer in the 1960s gets credit for the first iteration. But he brought the concept to new heights, lodging the number into the mainstream health lexicon. With his trackers, you could know every minute of the day how close you were to the daily goal. The number translates to roughly five miles—far more than researchers have since found important for improvements in either health or longevity. Park proudly talks about seeing people walk up and down airport terminals to get their steps in for the day, and a friend of mine once expressed great delight when I called her late in the evening.

"I'm only at 7,000 steps, so I can walk around my apartment while we talk!" she said enthusiastically.

Walking in circles around the living room is certainly better for your heart and mind than sitting on the sofa munching cookies. But even beyond the complete lack of research supporting the ten thousand steps fad, there's something unsettling about the whole trend. Instead of thinking about what their bodies actually need, the Fitbit aficionados rely on a computer to tell them how they feel—and it turns out that the computer data probably makes them feel worse rather than better. New research by Jordan Etkin, associate professor of marketing at Duke University, shows that people may get more exercise wearing a Fitbit—but they enjoy it less. The activity trackers and step-counting turn the simple pleasure of walking into an outcome-focused job. You're thinking about maximizing output rather than the immediate enjoyment of what you're doing. You stop enjoying the pleasure of the moment. Your happiness goes down, not up.

Etkin suggests that if you want to increase your happiness instead of just your steps, then throw away the trackers.

Measuring an enjoyable activity makes it feel like work. You focus on reaching an end goal rather than taking pleasure in any intrinsic sense of well-being. If you're preparing for a marathon or aiming for the Olympics, collecting data on your training sessions and movements is a great idea. For the rest of us, all those numbers may be a distraction from the hum of information that our bodies are providing on their own. When you're working *with* your body, you have a deep sense of pleasure. Mind and body are connected in one seamless system.

It's not just Fitbit. New technologies that supposedly put us more in touch with our bodies often do just the opposite, making us ever more separated from an instinctive understanding of our own needs. An Apple Watch uses sensors, data, and an algorithm to report whether or not you have slept well the previous night. It just never asks the basic question: *Are you tired?* Before lunch, you can check an app on your smartphone to tally how many calories you've consumed and how many more you should eat in the day. But understanding the telltale signs of hunger may be a better choice.

"Our feeling body is smarter than our thinking brain," says Dr. Judson Brewer, a mindfulness expert and psychiatrist at Brown University. "If we retrain ourselves to listen to the right signals, we won't overeat. Too often we lose touch with what our body actually needs and what it's trying to tell us."

Feeling the urge for a piece of cake could mean that you're hungry and your body needs food, but it could also mean that you're stressed, anxious, or bored and turning to the cake as a temporary balm to make you feel better. How do you tell the difference? "If we step back with awareness and pay attention, our bodies will tell us everything we need to know," Brewer says. "It's simple but not easy to listen to your body."

Simple but not easy. Our bodies are sending us all the signals we need, which is the simple part. But innumerable social, cultural, and psychological messages interfere, making it hard to hear them. When I spoke to Brewer, who also comes credentialed as a researcher and behavior addiction expert, he told me that if you have a craving, you shouldn't ignore it. Willpower usually fails in the long run because it's not how our brains and bodies work. Instead get curious and investigate your bodily sensations and what they may mean.

"Curiosity is a superpower," Brewer says. "Once you're curious about what's going on in your body, you're excited to get the information and learn more. Curiosity fosters awareness and jolts your brain out of autopilot. It moves better behaviors into the automatic mode."

You can take the same approach to any addiction—getting

curious about the messages your body is sending and pausing long enough to understand how they are affecting your actions. Brewer has used his techniques to help people stop smoking and tame the new demons of social media addiction. If you find yourself compulsively checking email or feeling the urge to text while you're driving, tap into your natural curiosity to understand what's happening in your body and mind. My own email habit had been getting a little out of control, and I agreed with Brewer that "you have to focus on the felt experience of behavior." So I gave the approach a try.

I had gotten into the habit of checking email on my computer dozens of times a day when I was supposed to be writing. If the email didn't prove distracting enough, I'd grab my phone and look at the weather app (which didn't change all that often) and scan the news headlines (also not urgent). Twenty minutes or so later, I'd remember to return to my writing—and I'd feel incredibly frustrated at the time lost.

The next time I got the urge to abandon the sentence I was writing for a quick email check, I sat back and thought about what I was feeling. My body seemed restless, and I was jumpy. I needed some distraction, and I realized that whether it was physical or mental didn't really matter. I stood up and walked around for about thirty seconds. When I sat down again, the urge had passed. I looked at my computer screen and managed to keep writing. For the rest of the day, every time I wanted to check my email, I asked myself *What is my body feeling?* I stood up a lot. I walked around. The time spent doing that was far shorter than the hours I would have wasted on the internet each time my brain needed a break. At the end of the day, I felt triumphant. I'd cut down my email distractions very significantly—and even more important, I had some sense of control back. If I got an urge to click or scroll or grab my phone, I could *study* the feeling instead of acting on it. I had been enormously skeptical of Brewer's idea, but putting it in action, I was won over.

What Your Neurons Say

Our bodies know more than we imagine, and they are telling us how to feel and behave even when we don't realize it. At his lab in New Haven, cognitive scientist John Bargh has been an important crusader researching embodied cognition—a relatively new field that is attracting psychologists, neurologists, and robotics developers. All of them are trying to understand the link between sensory input and the ways we think and what we experience. Some of Bargh's research findings are pretty stunning. Over and over, he and his colleagues find that the "embodied" part of ourselves—what our bodies are experiencing—plays a dramatic role in how we think, feel, and behave.

In one experiment that particularly amazed me, volunteers were handed a résumé on a clipboard and asked to evaluate the candidate. Imagine that you are one of the people getting the résumé. Since you know it's an experiment, you'd probably be on high alert. "Unconscious bias" has gotten huge attention in the last few years, so you might be extra careful that race and ethnicity don't affect your opinion. Many studies have shown that if you put a woman's name on a résumé, people are likely to think she's less competent than if you give the same qualifications with a guy's name, so you'd be sure not to fall into that trap. Age? Attractiveness? Where the candidate lives? By paying attention and making rational judgments, you could feel confident that you had evaluated the candidates fairly and without bias.

Except you'd be wrong—because it's unlikely that you would have considered the weight of the clipboard.

If you're now saying—*What? Come on! What difference could that possibly make!* I'm right with you. Who would ever think of that? Apparently Bargh did. Some of the clipboards were light and others heavy, and the weight of the clipboard changed how the candidate was evaluated. People who received the heavy clipboard rated the candidate as more serious and competent than people who got the résumé on a light clipboard. When I first read about the findings, I was gobsmacked. How could that be possible? The weight of the clipboard obviously has nothing to do with the person's competence—so understanding how it affects our judgment requires that we expand our sense of how our brains and bodies work.

Your body is constantly receiving input from millions of neurons that experience and interact with the environment. They then transmit messages to the brain, which can interpret them in many different ways. Why would a heavy or light clipboard matter? Consider that we think in physical metaphors. We often refer to topics like nuclear arms or global pollution as "weighty matters" or dismiss someone we don't respect as a "lightweight." Your unconscious mind spins and interprets the physical signal that comes from the heavy clipboard differently than when it's something light to hold.

It sounds pretty crazy until you realize that we are always taking in physical cues and using them as part of our attitude toward a situation. You can meet someone who seems to be your perfect (algorithmic) match on a dating app and fall hard for their witty texts, but if you don't like the way they smell when you finally meet in person, it's all over. Computer scientists are starting to recognize that even the biggest nerds in the world are more than disembodied brains—they have neurons that taste and touch and see and hear, and all of those are constantly contributing information that affects how they think. In other words, a lot of information goes from body to brain, rather than vice versa.

I got a firsthand experience of this pathway when I bought a new iPhone recently and something felt wrong. It seemed heavy and awkward, and using it didn't make me happy. I yearned for my previous five-year-old model. I went back to the store and a clever salesperson suggested that maybe the problem was the case I had purchased. The clear plastic had a hard surface with squared edges, and he suggested a soft silicone instead. The moment I tried it, the phone seemed transformed. Holding it was a pleasure—and wow, that camera was amazing, after all! He showed me another case with a silky texture. "The way the phone feels in your hand seems to transform the experience," he said. "I'm always surprised by how much difference it makes."

You would think that my brain would focus on the incredible technology in the very smart phone, but instead it got waylaid by the *feel* of the object. The soft case sent a comforting message to my brain that transformed what I thought not just of the case but of the actual function of the phone. The great cognitive linguist George Lakoff says that "our brains take their input from the rest of our bodies." Even when we try to think abstractly, we rely on physical realities and our own bodily experiences. Lakoff stunned people with his 1980 book *Metaphors We Live By* when he pointed out that our orientation in the world influences our most basic concepts about life. Human beings stand upright, our eyes at the top of our bodies, which leads us to envision everything good as being "up" and bad as "down." We'll say that someone is moving up in the world, or that a friend is in a downward spiral—and it doesn't need any further explanation. A very personal physical perspective even influences our understanding of religion. In imagining an afterlife, we glance up to the sky to indicate the empyreal place and downward for the devilish one. Rationally, we have no earthly (or heavenly) reason for that perspective, but if you try to envision good and bad from the opposite directions, you'll just feel confused. It doesn't merge with how our bodies perceive the world. "We cannot think just anything—only what our embodied brains permit," says Lakoff.

Take a look around the house or apartment where you live. Maybe you have soft furniture with down-filled sofas, cushy throw pillows, and comfortably upholstered chairs, or perhaps you prefer a harder-edged look with wooden Shaker furniture and solid ladder-back chairs. You probably think of each of these as an aesthetic choice that may even reflect who you are. But it also works the other way. Jonathan Ackerman, now a professor at the University of Michigan, led a study showing that people make different decisions when they're seated in a hard chair versus a soft one. The soft chairs made them more pliable in their attitudes toward others. Soft seat, soft heart, you might say. When asked to negotiate to buy a new car, those in the hard chairs offered dramatically less than the others after one offer was rejected. "Hard chairs made people harder negotiators," said Ackerman.

He calls this a "haptic mindset." I had to look that up. Haptic refers to our sense of touch, and with the clipboard and the hard chairs, he was able to show that what we hold or touch (even with our bottoms on the chair) unconsciously affects our mindset and attitude. There are very practical implications to the findings. "First impressions are likely to be influenced by one's tactile environment," he says. If you've ever shaken hands with a new acquaintance only to encounter a sweaty palm, you know exactly what he means. The sweaty-palm guy is going to have to do a lot of charming and impressing to overcome the don't-trust-him message that your body has immediately registered.

Out to dinner with friends recently, I described some of Ackerman's and Bargh's findings and mentioned that the more research I did, the more awed I became by the power of the body to influence the mind. Most people agreed, but one of the men at the table, a partner at a big law firm, just shook his head.

"It's all very cute, but my judgments could never be affected by clipboards or hard chairs," he said with (dare I say) just a trace of arrogance.

"The whole point is that you don't realize you're being influenced," I said.

"If you're rational, those things don't have an effect," he said adamantly.

I decided to let it go. We all like to think that we are in full control, making rational choices and decisions at every turn, and it's disquieting to discover that no matter how many advanced degrees you get, you can't outsmart your body.

The truth is that you don't want to outsmart your body. If you had to depend on your conscious mind to make every decision, you probably couldn't stay alive for an hour, never mind a day, month, or year. Reminding yourself to blink your eyes 960 times every hour and to breathe roughly as often would mean you couldn't do anything else. (And how would that whole breathing thing work when you were asleep?) When you're crossing the street and a car swerves in your direction, you really don't want to have to rely on your conscious mind to take care of things. By the time your brain can think—*My goodness, that Toyota is about to hit me. I'd better get out of the way. Let me just take a step back here*—you've long ago landed under the wheels. Fortunately, your physical reflexes function without any conscious thought. Only once you're back on the sidewalk does your mind get the message from your body—*Close call! But don't worry, I kept you safe.*

Neuroscientist David Eagleman points out that if you've ever watched a baseball game, you know that the body responds faster than the mind. When a pitcher throws a ball at 100 miles per hour it takes about four-tenths of a second to get to the batter. Conscious awareness takes about half a second—which means that the ball crosses the plate before the batter quite literally "knows" it. If the body weren't functioning without conscious input, nobody would ever hit a baseball.

"Your consciousness is like a tiny stowaway on a transatlantic steamship, taking credit for the journey without acknowledging the massive engineering underfoot," says Eagleman.

Interestingly, people like the lawyer who refuse to believe in the power of their bodies to influence their minds are the easiest for marketers and manufacturers to manipulate. When you're at a car dealership, for example, getting ready for a test drive, you probably think you're focusing on the car's safety, performance, and ease of driving. But car makers have figured out how to use your haptic mindset to their advantage, and many now put weights in the doors—recognizing that the sense of heft in your hand sends a message of solidity and reliability to your brain. My lawyer friend's refusal to believe that his physical sensations can influence his decision-making probably means that he will end up buying a car with a heavy door. In the long run, it's much better to relinquish a little of that sense that your mind is in complete control in order to gain more *actual* control. When you understand the signals your body sends, you can use them to make yourself happier and more productive.

You're Getting Warmer

Bargh's lab produced another mind-blowing experiment that reveals the power of the body to send signals to the brain even when we are completely unaware of it. In the study, a research assistant greeted each of the volunteers and took them individually to the lab. (Even she didn't know exactly what was being tested so she couldn't give anything away.) Along the way, she fumbled with a cup of coffee while trying to get some papers from her briefcase and asked the volunteer to hold the coffee. Once at the lab, the volunteers got to what seemed the heart of the experiment—they were given some forms with the description of an individual and then asked how much they liked them. But here was the twist they didn't know. The assistant had given some of them an iced coffee to hold, and some got a warm cup of coffee.

As incredible as it may sound, those who held the warm

coffee rated the person they read about as warmer and kinder. Those who had been primed with the iced coffee found the person colder and less likable.

Bargh's finding was one of those groundbreaking insights that seemed to send perceptions in a totally new direction. Some initial difficulty with replicating the results was later resolved, and researchers had a field day (as it were) looking for spin-offs. They found them. People who held the warm coffee become more generous when asked to give a donation than those primed with the cold cup. They show greater trust during a game where you have to depend on another person. They act more selflessly. Those are big results for a small cup of coffee. (Some of the experiments done in Europe used hot or iced tea rather than coffee, but of course it didn't matter—it's the temperature that made the difference, not the taste.)

"How do you explain it?" I asked Bargh, when we had a long and engrossing conversation by phone.

"It's fascinating," he said. "Some of it seems to be hardwired from ancestral heritage—it's human nature and universal. Other pathways are acquired in infancy. The baby is being held by a parent and held close, and the physical warmth gets conflated with being protected and taken care of. Warmth comes to stand for someone who is trustworthy and not a threat, the person who has your back."

Talking to Bargh, I could immediately imagine him in front of a classroom galvanizing students to think in unexpected ways. Years of research haven't dimmed his enthusiasm for the amazing connections between mind and body. He told me that he first got inspired to do the coffee experiment after watching a documentary on hell on the History Channel. It got him thinking about Dante's *Inferno*, the great epic poem that takes us through the nine circles of hell. If you're not up on fourteenth century verse, just know that each circle represents an increasingly terrible sin and describes the vile punishment that will ensue. Each delivers a *contrapasso*—a punishment that fits the crime. (Call it poetic justice.) The early circles are for lust, gluttony, greed, and wrath, and by the seventh circle we are at the place for murderers, warmongers, and tyrants, with lots of boiling blood and shooting flames engulfing the damned. Finally comes the ninth circle, the very worst, reserved for people who betrayed the close trust of others. Their punishment is to be frozen in ice.

"The poets grokked it!" said Bargh excitedly. "How amazing that in the midst of fiery hell, the worst punishment is being frozen. They intuited that a betrayal is like the loss of physical warmth. It changes everything."

Six centuries later, neuroscientists have finally given us a reason why Dante's ninth circle and Bargh's hot/cold coffee experiment both make sense. Using advanced imaging machines, they have identified a tiny area of the brain called the insula that gets activated when you're physically warm and also when you feel social warmth. We sometimes give excessive credence to the findings of neuroscientists (our new poets), but when the MRI machines, Dante, and a Yale psychologist are all telling you the same thing about the interweaving of mind and body, you definitely need to pay attention. Your body and your emotions are so deeply tied together that, in this case, it essentially requires only one control panel to handle both of them.

Taking these findings to the next step, two researchers at the University of Toronto wondered what would happen if they flipped the scenario. Would someone experiencing rejection—emotional coldness—seek physical warmth to make themselves feel better? To find out, they invited students to play a computer game where avatars tossed a ball back and forth among three players. (This game is a favorite among researchers—you'll hear about it again.) Some of the students got a few casual tosses at the beginning and then got ignored—essentially given the cold shoulder by an unfeeling computer algorithm. Afterward, when asked their preferences for different foods, they wanted hot coffee or hot soup far more than the students who had been included in the game.

We are talking here about an avatar and a random computer game. Yet even the virtual sense of being excluded (left out in the cold, you might say) changes what your body is telling you. It's as if your body knows that the hot soup will ease your emotional pain in a way that a cold Coke cannot. The Toronto researchers concluded that "experiencing the warmth of an object could reduce the negative experience of social exclusion."

It gets even crazier. When a young Dutch professor named Hans IJzerman had people play that same computer game, he found that, when they were excluded from the ball-tossing, their actual body temperature went down. In other words, craving the soup after a rejection had a physical cause that your mind simply never understood. A study at UCLA hospital recorded patients' body temperatures hourly and found they tracked to how close the patient felt to family and friends, rising when someone was emailing or texting or otherwise connecting to a loved one. Romance novels that pant about red-hot passion and feverish love have it right. Hot makes you hot.

The British idea of handling every crisis by serving a cuppa tea may have more than just tradition on its side. When you're rejected or feeling lonely, your body temperature drops about half a degree. You judge the room temperature to be colder than someone nearby who is feeling happy and connected. So go ahead and brew that pot of tea. A warm cup of something could be just the balm you need.

When my younger son was about ten years old, he wrapped his hands around the mug of hot chocolate he was drinking one afternoon and with a twinkle in his eye began sighing contentedly.

"Mmmmm, aaaaah," he said with obvious exaggeration. "Aaaaaah."

I asked what he was doing.

"Just copying you, Mom," he said with an impish smile.

I looked down and realized that I was indeed embracing my mug so the warmth of the porcelain penetrated the palms of my hands.

"You do that every time you drink tea," he said.

"I do?"

"Every time," he said.

It became a family joke after that—Mom didn't drink tea so much as cradle the cup for warmth. My younger son loved to imitate my wrapped palms and contented expression. None of us thought much about the *why* of the behavior—it was just one of my charming (maybe) quirks—but having talked to Bargh and studied the follow-up research on mind-body links relating to warmth and cold, I've started to wonder if the mugs served a deeper purpose. With two kids, a full-time job, writing, work travel, family events, and the other usual stressors of daily life at that time, I had a full emotional plate. Those moments holding the warm cup might have sent a physical signal to my brain of comfort and contentment. Everything was okay. My mind didn't have to figure out whether that was true or not—from some deep evolutionary grounding, it knew to accept it as fact.

I'm not the only one. In the popular TV show *The Big Bang Theory*, super nerd Sheldon Cooper is always offering distraught friends a hot beverage. "When a friend is upset, you offer them a hot beverage," he says in one episode. The writers might have meant this as a charming quirk that showed just how uncomfortable he was in normal social situations. But I like to think the writers had studied up on Bargh and IJzerman and used Sheldon's offers of tea as an inside joke about just how smart he really was.

Does all this research on warm beverages and soft chairs and metaphor of life hold up in real life and not just theoretical or lab settings? Drawing practical tips from research studies is always a tricky game (though journalists do it all the time) and there are always anecdotes to refute any study. You can no doubt tell me about someone whose Tribeca apartment is all raw-edged metal and wood but who's the softest person you know. Or how the only thing that makes you feel better when you're down is a Coke with lots of ice. We all have many inputs that affect our moods, attitudes, and decision-making. But too often we're not aware that our body is providing our brain with key information—and not the reverse. Recognizing that physical experiences are affecting you at every moment can go a long way to giving you more control as well as providing tools to make you happier at any time.

When I wrote *The Gratitude Diaries*, I discovered that if you're looking to improve your well-being, it helps to keep a gratitude journal by your bed. Focusing on the good things in your life makes them stand out and gives you a different perspective. I've been amazed by how well it works and can now suggest adding another little tidbit to the ritual. Have a sip of hot tea before you start writing. Your body being warm may help you think more kindly about other people and makes you more receptive to a grateful mood in general. If you really want to go the extra step, sit on a soft chair to write and get a journal with a smooth cover rather than a rough one. Giving body and mind a common experience of positive feelings can only help your spirit.

2

How Your Body Makes You Happy

> Sometimes your joy is the source of your smile, but sometimes your smile can be the source of your joy.
>
> —BUDDHIST MONK THICH NHAT HANH

AS I WALKED DOWN A street one afternoon, I caught a reflection of myself in a store window. Instead of the pleasant and friendly woman I expected to see, I glimpsed someone fretful and stoop-shouldered, racing down the block with her chin jutted forward, tension showing in every inch of her frame. Could that really be me? I didn't want to be late for my meeting, but I stopped in my tracks to regroup. An image flashed in my mind of the streaming show *The Marvelous Mrs. Maisel*, about a woman trying to succeed in the mostly male world of stand-up comedy in the late 1950s. Every time she goes on stage, her manager Susie mutters to her—"Tits up!" It's their funny shorthand for "stand up straight and be confident!" The thought made me smile. I pulled my shoulders back and lifted my head up and tucked in my chin. I focused on maintaining my uplifted posture as I walked down the street, and by the time I'd gone another block, the small smile had gotten bigger.

The meeting turned out to be fun with a lot of laughing on both sides. When I finally got back home, I tried to figure out if changing my posture had changed my mood—and whether that had improved my approach to the meeting. I concluded that the answer was yes and yes. Walking straight and with a smile, I'd felt less tense and unexpectedly confident. But I wondered exactly what the connection had been.

I got in touch with Erik Peper, a professor at San Francisco State University and an expert on biofeedback loops. He has won awards for his work in the field of psychophysiology, the connections of emotions and body, though he just calls it behavioral science. I asked him what he thought of my epiphany on the street about posture and mood. Given the extensive research he's done on the subject, he was completely on board.

"When there's a change in your body, your brain interprets it and applies a meaning," he told me. "We generally think of messages going from the brain downward, but a much higher percentage of fibers go from the body upward."

Understanding these pathways can go a long way to making you feel happier and more positive. Since the brain is often on the *receiving* end of the body's messages, Peper has looked at how posture can affect everything from mood to math skills. As we were talking, he asked me to try a little experiment. (You can do it along with me now.) He told me to sit hunched over with my shoulders drooped and my head down and to try to think of a negative experience I'd had. Almost immediately, an image came to mind of a day long ago when I'd been fired from a job and felt thoroughly stunned and defeated.

Next, he told me to try to remember a positive experience. Nothing quickly popped into my head. I was just grappling to come up with something when he told me to sit up straight and look forward—and try again. With barely any hesitation, I envisioned a glorious day strolling with my husband on the narrow streets of Venice.

"It's much easier to remember happy events when I'm sitting up straight," I said.

"That is the point," he said cheerfully. He's done this experiment often with large groups of students in his classes. In one reported study, he found that most people could recall positive experiences much more easily when they were seated upright and negative ones better when they were slouched and drooped.

"We all function on classical conditioning," Dr. Peper explained. "When we feel hopeless and depressed, our bodies collapse inward. Put your body into that position later and you evoke those same feelings. When you're upright, you still have access to the negative memories, but you feel more distant from them. You're more of an observer."

Being hunched over sends a message of defeat and hopelessness—and if the world is hearing that message, your own brain is incorporating it too. On the other side, think of how the winners at every track meet throw their arms up and out as they cross the finish line. One study found that even blind runners who had never seen anyone making that gesture did the same thing. It's instinctual. In victory, we stand tall and happily take up a bit of extra space. Defeat shows up as head hanging down and shoulders slumped.

For those who like to see hard data, Peper and his colleagues did one study where they wired up volunteers to EEG machines, which record electrical activity in the brain. They found a significant increase in "high-frequency oscillatory activities" when people tried to think positive thoughts from a slouched position. In other words, you have to work much harder to get a positive perspective if your body is drooping than if it's straight. You could think of that as the scientific substantiation of poet Maya Angelou's advice to "Stand up straight and realize who you are, that you tower over your circumstances."

The connection between posture and mood also makes sense on another level. I thought about George Lakoff's idea that our orientation affects how we see the world and considered my own experience on the street. When I was walking with my head thrust forward and down, my eyes stayed focused on the sidewalk. I wouldn't trip, but I also wouldn't take in anything around me. Once I had my head up and eyes forward, I had a more expansive view of the world. My senses were stimulated (very stimulated since I was walking near Times Square), and all the input unconsciously tweaked my perspective on the upcoming interview. When I was physically looking up, my mind made the leap that things were...looking up. I felt more charged and energized.

Your body position affects you in unexpected ways too. In another of Peper's studies, students were asked to subtract 7 from the number 964 and continue doing that serially for thirty seconds. If the very thought of that makes you want to run from the room, you're not alone. Sit up straighter, though, and it won't sound so overwhelming. When asked to rate the difficulty of the test, people who had high math anxiety rated it a 7 (out of 10) when they did the problems slouched over and only a 4.8 when they sat upright.

"That's a huge difference!" I said to him.

"Yes, but the biggest difference was people who already had math anxiety or test anxiety. For people who were already happy-go-lucky, the posture didn't make much difference."

If you're already feeling negative about the test (*I'm lousy at math!*), the slouched posture just reinforces the fear and makes it worse. Once your body sends your mind a more positive message, the new information makes you perk up. Now the neurons are firing the message—*This isn't so bad!* Your conscious mind incorporates the new information with the old fear and starts to even things out.

Once you start associating emotional states with your body position, you can very quickly start making yourself happier. For example, another link your brain learned long ago is that when you're physically threatened, your muscles tighten. Your brain associates tense muscles with danger and the need to be on high alert. Now imagine it's a Tuesday afternoon and you got held up at work and are worried about being late for kindergarten pickup. The anxiety makes you clench your muscles, and your brain kicks into its vigilant mode. You can try to talk yourself into a different mood by stopping for a moment of gratitude. *So grateful that I can leave work early...grateful I have this adorable child waiting for me!* It's a good start, but the positive approach conflicts with the story your body is telling.

As soon as you're aware of it, you can make a physical change that will send a new neural signal to the brain. Shrug your shoulders to release the muscular tension, and take a few deep breaths. With the muscle tightness eased, your body stops sending the danger message, and your brain withdraws its panicked alert. Now you really can appreciate that adorable child. Gratitude and happiness become much easier to find.

Your Brain Is Not a Computer

Recognizing the power that our bodies have to shape our thoughts and feelings changes what we can expect from some futuristic metaverse. In the virtual world, our bodies may simply cease to exist—and that is a serious problem. British intellect Jeannette Winterson believes that most of us have gone even farther than we realize in separating our minds and bodies. "Even now, because of the way technology is advanced, many people don't use their bodies at all in the real world," she says. "They use cars and planes to transport their brains around. They lie their bodies down and watch the TV. They sit in front of their laptop."

In his novel *Klara and the Sun*, Nobel prize winner Kazuo Ishiguro imagines a world where robots become the Artificial Friends (AF) of children. One AF named Klara gets taken into a family where the young daughter is ill, and the possibility is raised that if the child dies, Klara can replace her. The body is just a vehicle for consciousness—and if Klara understands everything about what the child says and does and thinks, maybe her corporeal form isn't necessary anyway.

This is more than a novelist's fantasies. In Silicon Valley, engineers talk about "embodied" intelligence—a human, a robot, or any form of intelligence that includes a material form. The other side of that is consciousness that exists disembodied. Artificial intelligence began with the idea that consciousness is all that matters and the form that Siri or Alexis takes is irrelevant. But that notion is being increasingly discredited as scientists across many disciplines discover that our bodies play a bigger role in who we are and our sense of pleasure and joy than we often understand. The newest breakthroughs in AI come from understanding that in order to be truly "smart," robots need to touch and experience and feel.

We can't completely blame technology for separating us from our bodies, since the conversation really began with Plato and Socrates (as so many do). Plato argued that mind and body are different substances, and he believed you need to trust your rational mind and not what your senses tell you. See that stick that looks bent when it's in the water? Beware drawing any conclusion! When you pull it out, it will be straight. Our modern experiences with airbrushed photos, edited videos, and rounds of "alternate facts" make his theories from the fifth century BC seem pretty relevant.

One of the biggest impediments to pleasure in our current times is our disconnection with our bodies and the world around us. If technology is causing some of the difficulties, it can also help solve the problem—giving us insights into just how important it is to have embodied intelligence. Rolf Pfeifer, who runs the artificial intelligence lab at the University of Zurich, points out that "brains have always developed in the context of a body that interacts with the world to survive. There is no algorithmic ether in which brains arise."

Pfeifer believes that the next generation of artificial intelligence requires figuring out how to let robots have actual physical experiences—something computer scientists now think is vital to any humanlike intelligence. The metaphor of our brains being like a supercomputer—getting input, processing it, then giving output—not only doesn't work for people, it's wrong for computers. The most cutting-edge roboticists now think that, instead of trying to program a computer to know what an apple tastes like, it's better if it can have the actual sensory experience. This whole new way of thinking about artificial intelligence also changes how we think of ourselves.

The Morning TV Trick

If you've ever interviewed for an important job or tried to ask a boss for a raise, you might have had the body-first experience. You think you're perfectly calm and well prepared, but the moment you try to talk, your heart begins beating hard and your throat constricts. You know what you want to say, but your voice comes out sounding funny and every part of your body is trembling. You didn't think you were nervous, but this is well beyond your conscious control. Your autonomic nervous system (which controls involuntary actions like breathing and blood pressure) perceived something big going on and sent out a surge of adrenaline. The physiological symptoms are instantaneous—and once your brain senses them, it's all over. Your mind picks up on the physical sensations and becomes as jangled as your body.

This syndrome happened to me early in my career when I started appearing regularly on morning TV shows. I prepared well and never felt nervous as I walked onto the set—but as soon as the stage manager said "We're live!" my autonomic nervous system made its own choices. (That's why it's called autonomic—it's completely involuntary.) However composed and unruffled I felt beforehand, my suddenly pounding heart and sweaty palms threw me off course every time. I tried deep breaths and calming exercises but nothing helped. Finally I got a prescription for a tiny dose of a beta-blocker drug, which stops the effects of adrenaline. I took a few milligrams thirty minutes before my next big show. It worked miraculously. I never had the problem again.

Beta-blockers turn out to be a well-known strategy for actors, musicians, and public speakers. For violinists who need extremely steady hands, they are the Stradivarius of drugs (though much more affordable). Here's the important point. They don't work by calming down your mind. They work on your body. We usually think that the feedback loop goes from mind to body, but this intervention makes it very clear that the loop also circles in the other direction. You can imagine a small part of your brain that is constantly scanning the input from your nervous systems and other physical states and trying to figure out *Everything good?* The mind doesn't provide the answer—the body does. With the body sending no signs of surging adrenaline like pounding heart and constricted muscles in my post–beta-blocker case, the mind sent out an *all clear* signal. I could do my TV interview or give my talk as planned. Learning about the body-mind loop wasn't just interesting for me—it helped save my career.

Botox for Happiness

At a recent Thanksgiving dinner, I encouraged everyone at the table to say one thing that made them grateful. It's a regular with our family and friends, so I knew everyone would be prepared. After many sweet comments about good health, new babies, and sublime adventures, a young friend named Cassie contributed an unexpected answer.

"I'm grateful for Botox," she said with a bright smile.

She didn't elaborate, and the next person jumped in with his own gratitude moment. As we cleared dishes after the soup, though, I cornered Cassie in the kitchen and asked her to explain. She had no lines on her face, but given that she had just turned forty, smooth skin seemed more like a normal occurrence than a source of gratitude.

"I'm grateful for Botox because I'm suddenly much happier," she said. "It's not why I got it—but it's what happened." It turned out that she'd had a skin cancer removed some months earlier, and the surgery had left a bump on her forehead that the dermatologist said would be smoothed out by a heavy dose of Botox. Shortly after getting the injection, Cassie noticed that not only had the bump diminished—her mood was different.

"I couldn't move my forehead, so I couldn't frown. I'm a very expressive person, but not being able to make a frowny face somehow changed how I felt. Anger disappeared faster. I felt much calmer."

She noticed a change in dealing with her mom too. While the two of them used to have highly emotional encounters, Cassie's new poker face seemed to change how her mom responded to her. With Cassie appearing calmer and less angry, her mother stayed calmer too.

"I know it sounds crazy, but I asked my dermatologist if

this could really be happening, and she said, 'Sure—there's even some good evidence,'" Cassie said.

To learn more about that evidence, I contacted Dr. Kenneth Arndt, who had been a professor of dermatology at Harvard Medical School and the head of the dermatology department at Beth Israel Deaconess Medical Center. Credential-wise, he is as respected and mainstream as any doctor could possibly be. I mention all that only because he co-wrote one of the original papers suggesting that Botox injections "might induce positive emotional states." Coming from someone else, it might have been dismissed as…slightly wacky.

When I spoke to Dr. Arndt recently, I asked if he really believes Botox might lift your mood as effectively as it lifts your brow.

"It's a real possibility," he said.

More than four million people every year use some version of Botox (it has other brand names now too) to smooth wrinkles in the forehead and around the eyes. It works by paralyzing certain facial muscles, making it harder to put your face into a frown. Botox has become such a cultural phenomenon that young movie stars use it and so do middle-aged women who get their injections in shopping malls. The botulinum toxin attacks and paralyzes nerves—which can have an obvious downside, including death. People used to worry about getting deadly botulism poisoning from damaged canned goods, but the same people who once threw away any tomato sauce can with the slightest dent now willingly pay hundreds or even thousands of dollars for shots of the toxin.

In the form used by dermatologists and other doctors, Botox is extremely safe. The cosmetic advantage is straightforward—you can't frown, so you can't get frown lines. But since the mind relies on feedback from the body's neural messages, a much more complicated loop can take place. If your face can't register negative expressions, it's possible that you won't *feel* those negative emotions either.

My friend Lynn Schnurnberger and I once wrote a funny novel called *The Botox Diaries*. The book got great reviews and we celebrated when it became (according to ABC-TV) the summer's must-have beach read. People understood that the title was a joking reference to women at any age who still want fun, sex, and new adventures. But now I realize that it had another resonance we never intended. Life is better if you don't frown. Your brain is less likely to register sadness and despair if you maintain a bright expression (however faked it may be).

Botox is relatively new, but the idea of linking of facial expressions and emotion goes back a long way. In 1872, Charles Darwin wrote that our expressions aren't just indicators of our emotions—they can actually *create* those emotions. If you want to know how someone feels, mimic their facial expression, and you'll feel the same way they do. Edgar Allan Poe came up with the same idea (in literary form) thirty years earlier. In the short story "The Purloined Letter," an eight-year-old boy who constantly wins at games tells the great detective Auguste Dupin that his trick is to copy the facial expressions of his opponents. He then understands who they are and how they'll behave. Dupin tries the same approach and he solves a case.

Pretty clever, right?

The great English philosopher Edmund Burke also understood the game back in 1757. He said that by copying someone else's face and gestures, you could understand their thoughts because you would then be feeling the same ones. For himself, Burke said, "I have often observed that, on mimicking the looks and gestures of angry, or placid, or frighted, or daring men, I have involuntarily found my mind turned to that passion whose appearance I endeavored to imitate."

That a great philosopher, an eight-year-old boy, and a fictional detective all made the same observation makes it intriguing—though not necessarily true. So a century after Darwin, researchers began looking for evidence that the emotions we show on our faces or copy from someone else can change how we feel. They discovered mirror neurons, which fire in the same way when you do a physical act and when you observe someone else performing the same act. Many studies link mirror neurons to understanding another person's goals or intentions. In that way, they seem to be involved in our ability to be empathetic—and possibly to solve murder mysteries.

At about the same time, neuroscientists proposed the "facial feedback hypothesis." In a nutshell, it says that the expressions you show on your face influence the emotions that you experience. True believers say it goes beyond influence and actually *determines* how you feel.

Why does it work? According to the research, it's all about the afferent neurons—the nerve fibers in our bodies responsible for bringing sensory information from the outside world to the brain. We know the basic senses of seeing, hearing, smelling, tasting, and touching, but the afferent neurons are also sending information about every stimulus in the body—from a change in temperature to a muscle twitch. The brain reads the stimuli from the muscles and then copies the emotion: *That's a smile. We're happy!*

The Smile Effect

A much-reported study by German social psychologist Fritz Strack sought to clarify just how powerful the effect of facial expressions may be. Participants were asked to hold a pen between their teeth in a way that would make their muscles contort into the equivalent of either a smile or a frown. They then judged how funny they found a series of cartoons. It's important to note that the participants weren't asked to fix their faces into anything that had an emotional meaning. They didn't consider that they were smiling or frowning—they just had to focus on holding a pencil. The simple muscle movements that ensued, though, sent the signal to the brain, which is constantly scanning muscle stimuli to determine how to feel. The people whose muscles were in a smiling position found the cartoons funnier.

The study hit some controversy over replication (a common problem of late), and for a while it looked like Strack's conclusion might be discredited. But not so. In the years since he first published the study in 1988, researchers have done similar experiments over and over again with different tweaks and different results. (I can only begin to guess how many PhDs have now been granted based on follow-up research to Strack's.) Collecting data from 138 studies, researchers at the University of Tennessee and Texas A&M concluded that making yourself smile really does improve your mood. Looking for even more certainty, researchers in labs around the world have now united to share their data. A young researcher at Stanford named Nicholas Coles helped organize the multi-lab experiments. He says that the first pilot studies showed that "smiling could both magnify ongoing feelings of happiness and initiate feelings of happiness in otherwise non-emotional scenarios."

If you're starting to think of reasons other than the facial feedback hypothesis to account for why putting your face into a smile increases happiness, the researchers are a step ahead of you. Maybe it's a placebo effect? Once you expect that smiling will make you happier, it does. Great thought—but Coles reports that the results of six experiments done in twenty-nine countries show that's not the case. Even when people were explicitly told that facial poses would *not* affect emotions, they did.

Growing up, I sometimes walked around my parents' house with the dour, put-upon expression of a typical suburban teenager. My father referred to it as "the look." He would put his face close to mine, and giving an exaggerated grin, he'd insist "You need to S-M-I-L-E!" Needless to say, the suggestion never had much effect. In fact, it annoyed me. I'd read enough feminist literature to know that women over the centuries had been taught to smile as a way to show themselves as pleasant and submissive. Beauty queens and geishas smiled. Angsty intellectuals and Nobel prize winners didn't have to. All these decades later, though, I have a new view of my dad's advice. He simply wanted me to be happy—and he instinctively understood that smiling might help. He loved the Nat King Cole song about smiling through your fear and sorrow, and he took it as a lesson for how to make yourself happier.

After I was long gone from the house and my dad had gotten older, he often played Cole's "Smile" as an inspiration in his own life. By then everyone from Eric Clapton to Barbra Streisand had recorded a version, and more recently, Lady Gaga performed it. The song doesn't just say that things will get better—it gives you a way to break through the clouds. Smile and you will start to feel like the sun is shining again.

Neither my dad nor Nat King Cole had any suggestions on *how* to smile, but the model Tyra Banks came up with the technique of "smizing"—which means smiling with your eyes as well as your mouth. She taught the method to the young women on her TV show *America's Next Top Model* in the early 2000s, and it got a surge in attention during the covid pandemic when mask-wearing became the norm. She explained that the "right smize" comes from thinking of "something that delights you."

Smizing is more formally known as the Duchenne smile, after the nineteenth century French physician (Guillaume-Benjamin-Amand Duchenne de Boulogne, if you want to be formal) who used electrical currents to stimulate and study muscles. He concluded that only a smile that involves the *orbicularis oculi*, the muscles that surround the eye socket and squeeze the corners into crow's feet, shows true enjoyment. Some research shows that once you learn how to fake *that* smile (people used to say you couldn't, but you can), your mood will improve even more dramatically.

All this brings us back around to Botox and the contention that Ken Arndt and his colleagues made in an early paper that giving Botox to reduce facial lines could have the side effect of "reducing the internal experience of negative emotions" and making patients feel "less angry, sad, and fearful." As far as drug side effects go, those are pretty good.

You can understand why Botox got the interest of the neurologists trying to understand our mind-body connections. A young German researcher named Andreas Hennenlotter wondered what happens to the neural activity in the brain when you get Botox. To find out, he used the now-popular technique of putting people into MRI machines and scanning their brains under various conditions. He asked people to imitate angry expressions, and he discovered that when the muscles couldn't contract into a frown because of Botox, a part of the brain called the left amygdala didn't activate as strongly. If you need a good excuse to yourself for getting Botox that doesn't involve vanity, this may be it. After Botox injections, the brain registers less intense feelings of anger and negative emotions.

As a logical next step, Dr. Hennenlotter considered the possibility that Botox injections might treat mild depression, and in a small study, it relieved symptoms 90 percent of the time. While the statistic seems excessively high, the actual concept is standing up to scrutiny. Dr. Arndt reports that in studies of patients with major depression, about 50 percent are improved after Botox injections. Patients in a control group who get a saline injection improve about 15 percent of the time. (That's the power of the placebo effect.) It's early days in the Botox-depression research, but the connection seems to be there. At the risk of sounding like either my dad or Nat King Cole, try smiling when your heart is breaking. You might trick your brain into feeling happier.

3

Your Mixed-Up Mind

Whether you take the doughnut hole as a blank space or as an entity unto itself is a purely metaphysical question and does not affect the taste of the doughnut one bit.

—HARUKI MURAKAMI

NOT UNDERSTANDING THE FULL CIRCLE of body-mind connections can lead to all sorts of counterproductive activities. For example, if you visit Las Vegas and have $55 to toss away, you could put down a big bet on a roulette wheel—or you could use it to buy ten minutes at the Rage Room in Sin City Smash. After putting on protective gear including a safety helmet and coveralls, you're given a sledgehammer or bat and led to a room filled with items you can smash. Outdated electronics are often included (finally a use for old VHS players) as are dishes and glasses. Ten minutes of smashing is supposed to help you release your pent-up anger and stress. If you're really angry, you can get the $95 premium package, which lasts twenty-five minutes and assumedly has more recent computer monitors and nicer dishes to destroy.

Far crazier things take place in Las Vegas than a room where you break stuff. But as with other Las Vegas activities, this one can backfire and what starts out as fun can become more dangerous. The problem is that a Rage Room is more likely to increase your rage, anger, and negative feelings than to release them. One study out of Iowa University found that hitting a punching bag made people angrier and more aggressive, not less. Once you understand that the body-mind connection works in both directions, the result makes total sense. When you pound a pillow, scream into a towel, or smash china, your body sends out the chemical signals that come with danger, anger, and aggression. Your brain picks them up, and suddenly you really *are* angry or excessively aggressive.

Sin City Smash promotes itself as entertainment, as do most of the Rage Rooms now appearing in chic spots like Brooklyn, New York, and Austin, Texas. Go to one of them with a bunch of work colleagues for a few beers and a good laugh, and no harm is done. But don't expect to feel relaxed and de-stressed when you leave, because the physiological response will be quite the opposite—and you'll feel more combative and revved up.

How did we possibly get the idea that making your body enraged would put your mind into repose? Physiologically, it simply doesn't make much sense. The concept can be traced back to Freud (who got so many other things wrong too). He believed that when negative feelings build up, they produce a pressure that needs to be released—much as happens with pressure cookers or hydraulic engines. The release is called catharsis. If you don't vent the pressure, it will explode.

The problem is that people are not teapots who need to let off steam or else blow a gasket. One psychologist disdainfully told me that Freud took the children's ditty "I'm a Little Teapot" a bit too literally. Over the last fifty years or so, psychologists testing the catharsis theory have found almost no support for Freud's idea. In fact, they have discovered quite the opposite. As Brad Bushman, a chaired professor at Ohio State University concluded after analyzing decades of research, "venting to reduce anger is like using gasoline to put out a fire—it only feeds the flame."

Screaming Makes You Angrier

Since Bushman is one of the great experts on the causes of aggression and violence, I was excited to talk to him about the Freudian confusions. He laughed when I mentioned his reputation as a myth buster, but he quickly concurred. "I like to bust myths people hold that contradict the scientific evidence," he told me.

If you think that punching a pillow or screaming into a towel will help you get over angry feelings, Bushman has the evidence to prove otherwise. He points out that anger is characterized by high levels of arousal. When you're upset, your blood pressure goes up and your heart rate increases. "If you want to reduce anger, you don't want arousal levels to get higher," he says. Yelling, kicking, screaming, or hitting a punching bag (even in a fancy gym) just adds to the physiological arousal.

In one large study, Bushman divided people into three groups. He started by getting everyone angry at another (made up) person who had criticized them. One group then got to hit a punching bag, and they were told to think of the person who had criticized them. *Imagine his face on the punching bag! Get out your anger!* The second group also got to attack the punching bag, but they were told to think of the activity as physical fitness. The last group sat quietly—no hitting or banging or screaming. After that, everyone got to confront the person they thought had angered them. Bushman measured their levels of aggression by letting them blast their adversary with high levels of noise through headphones (a better choice than knives and guns in a lab). The result? The people who walloped the punching bag under either condition were much *more* aggressive than the ones who had sat quietly. Far from making them less angry, the venting got them even more revved up.

In another study Bushman and his colleagues gave people an article to read that had supposedly appeared in the well-respected magazine *Science*, reporting on a study done at Harvard University showing that venting anger works. All of it was bogus—but they wanted to check out the placebo effect. If you think something will change your attitude, will it?

"We were successful in manipulating people's beliefs," Bushman told me, "but even when people thought venting would work, it didn't." Even though they expected that the punching bag would dispel their anger, the people who tried the fisticuffs were still the most aggressive and angry afterward. "Call it the anti-placebo effect—and that should be the final nail in the coffin of the venting theory. If it doesn't work even when people *believe* it will, you have pretty damning evidence."

The very compelling (and damning) evidence should be enough to leave the venting theory in the dust. But deep-seated ideas are hard to shake. Since the 1980s, a company called Wham It! has sold (among other things) a forty-inch-tall inflatable bag that advertises itself as a cure for frustration that you hit to "get rid of all your stress and anger issues once and for all." Various pop psychologies offer recommendations like putting a photo on a pillow of an ex-spouse who annoys you and smashing it with a foam bat. A friend of mine who sees an expensive Upper East Side therapist to deal with decades-old anger at his father was told to think of his parent while punching into the air over and over. He took great umbrage when I expressed doubt about the technique.

"It worked," he insisted. Then lowering his voice, he added, "As I was punching the air, I started to cry."

I had no doubt that the action made him emotional—but probably for the opposite reason than he thought. He wasn't releasing his anger at his dad—he was just repeating it.

"When people vent, they tend to ruminate on whatever made them angry," Bushman told me, "and a large body of research shows that rumination is a horrible strategy. You don't want to rehearse what made you angry. Punching the air and thinking over and over about how you've been wronged is a terrible idea."

Why do people keep on venting—despite all the evidence that what's happening when you yell or scream or hit someone is the opposite of catharsis? "It feels good in the moment," Bushman said, "but chocolate and street drugs can feel good too, and none of these are good solutions."

So what does work? What could I tell my friend to try instead of punching the air? Bushman's impressive research had me convinced that trying to physically release your anger is completely counterproductive—so maybe the best technique for happiness is just to ignore the feelings altogether. I mentioned to Bushman Charles Darwin's idea that expressing an emotion intensifies it while repressing a feeling makes it much less potent. Could it be that repressing our feelings has gotten a bad rap over the years? Maybe the Brits got it right with their stiff upper lip and we should all just Keep Calm and Carry On.

Bushman listened politely to my idea, but he didn't agree. (Probably a good thing for me since I have no personal skills at repression. I am a blurter, not a concealer.) In terms of lowering negative emotions, repression is better than venting, but neither is particularly good for your heart or your spirit. He had a different solution.

"If you go back to the analogy about a teapot on a stove and the pressure building up, it seems like you have two choices—to vent or hold the steam inside. But people forget about the third option, which is to turn down the heat on the stove."

When you're angry and start yelling, you just get angrier. Your tense and ready-to-strike muscles send signals to the brain, which responds by revving up the response. It's not unlike two-year-olds having temper tantrums—as they kick their legs and pound their fists, their little bodies just get more and more furious. Trying to talk them out of their red-faced state isn't going to work, but if you swoop them up and hold them in a tight hug, they'll usually start to calm down. The quieting body tells the overcharged brain, "You know, things aren't so bad, after all."

Want to be happier when you're feeling angry? The secret is to lower your physical arousal levels. Your emotions will calm down too. The old nugget of counting to ten before responding works because as time passes, arousal levels naturally decrease. The heat gets turned down. You could improve your mood by sticking with Thomas Jefferson's famous advice to count to 10 when you're angry and 100 when you're very angry. Bushman says that when you're feeling angrily aggressive, you naturally lean forward, so to reverse the feeling, take a different stance. "Lean back in your chair and take a deep breath," Bushman suggested. Also on his list are any distractions that will divert your attention and help lower your physical agitation: Pet a puppy, take a warm bath, go for a walk, light a candle, do a crossword puzzle.

As for smashing china in a Rage Room? If you want to get flushed, sweaty, and increasingly aggressive, go for it. If you're looking to feel better and happier, consider a stroll through Pottery Barn instead.

Understanding Your Mixed-Up Mind

Our brains may be smart, but they sometimes mess up on interpreting the body's signals. I thought about that as I lay in bed the other morning, unexpectedly dejected and not wanting to get up. What was making me so morose? Nothing seemed particularly wrong in my own life, and I began thinking despondently about the bigger problems of the world. With a sigh, I rolled over and got a tissue to wipe my red and weepy eyes. I hadn't been crying, but I'd gotten an eye infection the previous day. Tears had leaked onto my now-soggy pillow all night.

With the tissue halfway to my face, I suddenly paused. Could my weeping eyes have anything to do with my desolate mood? The tears had come from bacteria, not emotion, but from a physical perspective, I'd been crying for hours. My brain had picked up that neurological information and sent a message of being down in the dumps. I'd like to think my mind is savvier than that—but I'd probably be wrong. "All emotions use the body as their theater," says the great neurologist Antonio Damasio. Almost as soon as I started thinking about my low spirits in terms of the weepy eyes, my mindset completely changed. The world didn't look quite so bad.

When some outside event causes a change in your physical state, your mind gets busy making sense of it. The problem is that physical symptoms don't come with instruction manuals. Red and weepy eyes could mean sadness or depression—or it could be that you're cutting onions for a casserole. Looking at any constellation of symptoms, a well-trained doctor immediately considers a variety of potential causes and goes with the most likely. Your brain does the same. It should be able to figure out the onions problem, but other distinctions are less clear. If you have a pounding heart, sweaty palms, and rapid breathing, it could mean you're sexually aroused by the person next to you—or it could be that you're just plain scared. How to tease out one from the other? The symptoms of surging adrenaline are so similar that our brains can easily misattribute the source. This misattribution of emotion happens more often than we realize.

In one famous study, researchers had men walk across a swaying suspension bridge. Halfway across, a woman stopped to talk to them under an experimental pretext and provided her phone number. About half the men on the suspension bridge called her afterward. Were they attracted to her? Well, maybe. But it's also possible that as they stood talking to the woman, they attributed their pounding hearts to her. Rather than admitting to themselves that the bridge crossing had them scared, they attributed the physical signs they felt to her presence. As a control, the researchers had other men meet the woman as they walked across a solid bridge. Since it wasn't scary, the men had no adrenaline surge to cause the pounding heart, sweaty palms, or shortness of breath that the suspension-bridge guys felt. Only about 20 percent of those on the solid bridge did the follow-up call.

Why Roller Coasters Are Sexy

With the insight from the bridge experiment, cheesy magazines began advising a roller coaster ride as an ideal place to take a first date. A potential partner experiencing heart-pounding, wide-eyed arousal might confuse the physical signals of fear and think "Very attracted!" Six Flags surely benefited. I have no doubt some researcher will eventually look at how many sexual liaisons occur after participants plunge 255 feet at 70 miles per hour on the Goliath roller coaster in California. A similar pattern may even explain why horror movies like *Scream, Halloween,* and *Nightmare on Elm Street* are popular for date nights—and have so many sequels.

This all sounds rather sleazy now that we have a new perspective on gender relationships, and hopefully we can use the understanding of misattribution for more positive results than surreptitious seduction. But it doesn't change the fact that our bodies are always sending the signals that our brains interpret, correctly or not.

The roller coaster and *Scream* movies give support to what

psychologists Stanley Schacter and Jerome Singer dubbed the two-factor theory of emotion. Factor #1 is a physiological arousal. Your body gets revved up, and the intensity of the revving determines the intensity of the emotion. Defining the emotion comes with Factor #2. Your cognition kicks in and tries to figure out *What the heck is causing all this arousal?* Your brain scans the sensory input and physical responses and tries to get some context.

If your heart is pounding as you wait to get test results in a doctor's office, your brain makes one guess. The same pounding when your naked lover walks into the room brings a different conclusion. In an early experiment testing this theory, Schacter and Singer gave people shots of adrenaline without letting them know what to expect, then introduced them to someone whose behavior was either euphoric or angry. Factor #2 kicked in, and the participants quickly concluded that they too were euphoric or angry.

The research and the swaying bridge reminded me of a conversation I had with a roommate when I first moved to New York. She had come to the city from a small town in the Midwest to become a model, but far from being naive, she had a natural savvy. After a guy she had started dating left late one night, I asked her how their evening had been.

"He told me he loved me," she said.

"That sounds serious," I said.

She rolled her eyes. "Oh please. We'd been lying in bed and he was excited. That's not a time to believe 'I love you.'"

We both laughed. Maybe she wasn't a romantic, but she was wise. Your heart is pounding when you're sexually excited and your physical arousal is at a peak. Your brain searches desperately for context, and in that moment, the perspective can easily go astray. Love? Passion? Sexual desire? New information in a less aroused moment (*I just wanted a one-night stand*) may change the evaluation. My roommate joked that she planned to needlepoint a slogan for over her bed: "Don't trust *I love you* from a man who has his hand on your breast." I think it should be a big seller.

The Positive Side of Misinterpreting

I recently saw a popular gif of the great actress Lucille Ball going from wild laughter to sobbing tears in a couple of seconds. It's thoroughly engaging. Her expressive face animates both the joy and the distress—and the flow from one to the other appears seamless. Part of what's so funny—and also what makes it very realistic—is the extreme of both emotions. You've probably been with someone who gets overcome with emotion and, as her shoulders shake and tears gush, says, "I don't know if I'm laughing or crying." How can you not know? In terms of physiological arousal, the two are very similar. Your mind needs circumstances and context and environment to separate them, and even then the line can be blurry. A friend of mine says that whenever she goes to a wedding, she gets choked with emotion during the ceremony. A lump rises in her throat so she can barely talk, and tears sting the back of her eyes. "I sit there wondering if I'm overcome with joy for the couple's future—or full of despair that it's not going to be as wonderful as they think."

Joy or despair? Your mind takes the input from your bodily sensations and the environment and then decides what the interpretation should be. And that means you have enormous power to change your emotions. When my friend gets choked up at the wedding, she can decide that she is feeling the sadness that her own marriage isn't quite as perfect as she dreamed on her own wedding day. Or she can choose to take those same bodily sensations and turn them to a happier, more upbeat view—deciding that she is warm-hearted and joyous and filled with the glow of love from the young couple.

"Emotions are your brain's best guess of what your bodily sensations mean," says Lisa Feldman Barrett, a professor of psychology at Northeastern University who focuses on how emotions get made. "Emotions feel like uncontrollable reactions that happen to you, when emotions are actually *made* by you."

Your body sends the signal, and your unconscious mind gets a shot at deciphering it first. But you're not at the mercy of those firing neurons. You can redirect how they're interpreted. Understanding how to reappraise your body's signals can do a lot more than make you happier at weddings. It can change everything from the likelihood that you'll get a raise at work to your success at karaoke.

If you want to change how you interpret an emotion, you

need to think about the message your body is already sending. Let's say you're about to give a speech, meet with a boss, or go on a first date. Your body recognizes a challenge ahead and shoots out the hormones that rev your body for action. Experiencing a pounding heart, sweaty palms, and fast breathing, you probably tell yourself *Calm down!*

Good idea—but there's a problem. The physical symptoms you're experiencing are definitely not present when you're calm. Asking your mind to fight against all the information your body sends is unlikely to lead to much success. A much better idea is to start with your physical state and make sure that the interpretation aligns with it.

Since the mind can sometimes misinterpret what the body is saying, why not give it a little nudge in a positive way? Alison Wood Brooks, a professor at Harvard Business School, had a clever idea for how this might work in the real world. Two emotions that we all feel quite regularly—anxiety and excitement—feel very similar in your body. Both put you in a state of adrenaline-induced high arousal. If you're about to go into a meeting with your boss, your mind reads the situation and sends you into a spiral of anxiety. Brooks suggests making a simple switch. Instead of

I'm so nervous about this meeting!

tell yourself that the message is

I'm so excited about this meeting!

Suddenly the same signs of surging adrenaline feel good and desirable rather than upsetting.

Does it work?

In a string of experiments, Brooks put people in situations that might make anyone tense, including giving a speech and taking a math test. But the one that makes me tremble just to hear about it involved karaoke—specifically, singing the first verse of "Don't Stop Believin'" by Journey in front of other people. (I have a notably terrible singing voice, so the $5 each participant got doesn't seem nearly enough.) She had some of the participants say "I'm anxious" ahead of time while others said "I'm excited." There was also a neutral condition in which people said nothing. Then they were taken into another room and given a microphone to perform the karaoke. (Forgive me, my heart rate is rising even as I write this.) When they finished, the program's software gave an objective performance score.

It's an extremely simple intervention—and yet it worked. The people who had said "I'm excited" got higher scores and reported feeling better about themselves. In other parts of the study involving public speaking and a math test, the same results occurred. The "I'm excited" group were judged by independent evaluators as better and more successful than the others. They also felt better about themselves.

Brooks expects that the anxiety-to-excitement technique works because both emotions are what she called "arousal congruent"—meaning your body feels pretty much the same way for both. But the results are completely opposite. Anxiety makes you feel generally awful and interferes with a good performance. Excitement is positive and pleasant and helps you do better.

"The way we verbalize and think about our feelings helps to construct the way we actually feel," Brooks concluded.

The results were published in the Journal of the American Psychological Association—and the research seems impressively solid. Brooks thinks that the more times you try this technique, the happier you'll be. I expect she's right.

The Sunny Side of Stress

When I wrote *The Gratitude Diaries*, I learned the power of reframing situations. You can't always control the events that happen in your life—but you can control how you think about them. No matter how bad the problem, you can always make yourself stop and look for the good side. My husband seems to be a natural at this. The other night he got caught in traffic for more than an hour and a half on his way home, long enough to make most people go thoroughly bonkers.

"You must be so frustrated," I said when he finally came in the door.

"Not really," he said. He put down his briefcase and kissed me on the cheek. "I'm listening to a really good book in the car. I'm grateful I got in an extra chapter or two."

Reframing doesn't make the bad stuff go away, but it gives you a different way of thinking about what has happened. You can try reframing in almost any situation. When I'm frustrated or angry, I sometimes make myself physically stop and think *What's the good side of this?* I stay right where I am until I can find a sliver of brightness to lighten my mood.

The anxiety-excitement experiments show that a version of reframing (in that case, renaming) works with emotions too. In the same way that my husband reframed his evening commute in traffic from an exhausting drive to a great time to listen to a book, you can get a new perspective on the emotions your body is fomenting.

Emotional reframing can have surprising and long-lasting effects. In one study, people who were about to take the GRE, the standardized test that's used for graduate school admissions, were invited to a lab to take a practice exam. Test-taking makes just about anyone nervous, and when it's a national test where (it seems) your whole future rests on your doing well, nerves can be particularly high. A lot of research shows that anxiety undermines performance, but in this case, as the students sat down at their computers, some of them were told that physical arousal actually *improves* performance. Their nervous energy was just their body mobilizing to do its best. The anxiety they were feeling wouldn't hurt their score and could help them do well.

The result? The students who got the anxiety-is-positive pep talk did dramatically better on the practice GRE. Not only that, when they took the real GRE a month or two later, their scores remained higher. Jeremy Jamieson, who led the study, calls this process "reappraisal." You recognize the anxiety but expect it to bring a different outcome. And surprisingly, the expectation pays off.

Jamieson did the study when he was a postdoc at Harvard in psychophysiology. (Hooray that the study of mind-body connections has Ivy League credibility!)

Now an associate professor at Rochester University, he shrugged off my compliments about the GRE study when we spoke, excited instead to talk about a paper he published recently in *Nature* that takes the stress-reframing idea even further. He and his colleagues worked with students in a range of demographics, ethnicities, and socioeconomic backgrounds to consider the long-range effects of thinking about stress in a new—and positive—light.

"People have come to believe that stress is bad and that if you feel your body having a stress response, you should do everything possible to get rid of it," he says. "But there are different kinds of stress, and we want to teach people to see stress as a helpful resource."

Stress as helpful? It's not a perspective you hear very much—or at all. The stress response evolved long ago to protect us (or our ancestors) from physical dangers. For the last few decades, researchers have said that the hormonal changes that occur at times of stress can be counterproductive if, instead of facing sudden physical challenges, you're confronting daily emotional tensions. The surging adrenaline and increase in cortisol that kick in might help you survive a charging bison, but they're not so great if the stressful event is going to a dinner party where you don't know anyone. In both cases, for example, the stress response will keep blood in your body's core rather than letting it stream to the extremities. (Yes, when you're scared, you literally have cold feet.) The response might keep you from bleeding to death if that bison attacks, but all it will do at the dinner party is make you want furry slippers.

Jamieson says that the received wisdom about the stress response being incompatible with modern problems may need a re-think. Or a reframing.

"The stress response is really a set of tools for dealing with difficult challenges," he says. "Signs of arousal aren't negative. Bodily changes aren't harmful. It's all in how you appraise them." Putting a positive spin on the physiological stress response means realizing that "your racing heart is sending more oxygenated blood to the brain so you can process things faster. Your whole body is mobilized to be more energetic."

Part of the brilliance of this approach is that your body and mind are no longer working against each other. Instead, they're both gearing you up to succeed. "Stress is about engagement. You wouldn't experience stress if you didn't care," he says. You may need to be in a high arousal state to do your best. The general recommendation to step back, take a break, and not try so hard when you're stressed "may not be the best advice. Sometimes you need to do hard things."

The psychologist Carol Dweck did pioneering research on the difference between what she labeled a fixed mindset and a growth mindset. When you have a fixed view of your abilities, you generally use all your experiences as reinforcing proof that you are (or aren't) smart or talented. A growth mindset sees every test or challenge as a chance to improve. People with a growth mindset get engaged by the process because they're less afraid of failure—and so the meaning of effort and difficulty gets transformed. Instead of feeling dumb when you don't know something, you get excited. Your neurons can make new connections. You can get smarter.

Jamieson and his colleagues realized the synergies between the stress-is-good approach and the growth mindset. A challenge is more exciting and manageable if you can see what's happening in your aroused body as exciting and manageable too. "You can use stress as a tool, rather than something you need to get rid of," he says. It worked impressively in his experiments, but I wondered how well that might translate to a daily experience. I asked how your revved-up body might be helpful when you're about to give a toast at a wedding—a common terror for any best friend asked to stand with a champagne glass.

Instead of thinking that you're nervous to be speaking in front of so many people, Jamieson says to "remind yourself that your body feels this way because the couple means a lot to you and you care. Your speech will then convey that sense of caring—and it's much better than being so calm that you come across as a robot."

You can try the same approach if you're asking your boss

for a raise. Remind yourself that you're nervous (with pounding heart, et al.) not because you're afraid they'll say no but because you want to push forward and take on harder things. Your body is getting you ready for that—and your passion and energy are all positive.

A few days after I spoke to Jamieson, I gave a talk about gratitude to an audience of four hundred people. Just as I walked on stage, I remembered that I had forgotten to take my usual beta-blocker to tamp down the physical signs of stress. (Or maybe my subconscious had purposely forgotten.) It didn't matter—I gave a big smile and felt fully in control. I began to talk, and just a couple of minutes in, my heart began pounding unexpectedly and I felt suddenly hot and sweaty.

In previous times, the message in the back of my mind would be *Your voice is going to shake and you'll be doomed!* But now I knew that the symptoms would pass in a minute. The tape playing in my head was *You're excited and ready to do great!* Sure enough, in under a minute, the symptoms subsided, and I continued with unusual energy. Jamieson is now doing some research to see if beta-blockers can actually make performances, notably by singers and some musicians, flatter and less engaging. Early results suggest that audiences pick up on emotions, and a bit of positive stress can make you a more engaging performer.

The speech I gave that day was one of the best I've done in ages. Maybe my mind and body really were working together. The thought makes me…excited.

4

How Your Senses Give You Joy

> Lean your body forward slightly to support the guitar against your chest, for the poetry of the music should resound in your heart.
>
> —ANDREAS SEGOVIA

A NEIGHBOR CAME TO MY front door recently with a big loaf of freshly baked bread. She had just put her house on the market, and the real-estate broker suggested she get busy in the kitchen. A maxim among Realtors has long been that a comforting smell wafting in the air will make potential buyers feel happily at home and encourage them to put in a bid. It seemed like a lot to expect from flour and sugar, so I thanked my neighbor for the bread and began to do a little research.

I quickly discovered that smells do indeed have a powerful effect on moods and emotions. Venkatesh Murthy, chair of the department of molecular and cellular biology at Harvard, says that smells make their way almost instantly to the limbic system, the part of the brain involved in emotion and memory.

That's why scents are so evocative. You may unexpectedly find yourself thinking of your college boyfriend some afternoon and then realize that a guy nearby is wearing the same cologne. Opening a box of old toys in my attic recently, I caught a whiff of the stuffed Elmo that my older son used to clutch tightly as a toddler. The pungent smell evoked such happy memories of my then two-year-old boy that I couldn't bear to throw beloved Elmo away.

The structure of the brain explains why memory and smell are so strongly connected. Elmo's scent landed squarely in my hippocampus, where memories are also stored. Before I had any conscious thought about it, the scent and the memory had mingled. Rachel Herz, a psychologist associated with Brown University, says that these kinds of smells can "increase positive emotions and decrease negative mood states." Anyone else might have found Elmo excessively rank, but to me he evoked childhood joy and sunshine—and he lightened my mood all day.

The connection to a happy memory is one reason a smell can unexpectedly (and unconsciously) make you feel grateful or content—but it's not the only one. Consider, for example, a study out of Australia that found the smell of fresh-cut grass makes people feel more relaxed and joyful. From our knowledge about brain geography, you might guess that one whiff of that newly mowed grass links instantly to memories of romping joyfully outside. Reasonable enough—but it turns out to be more complicated. The neuroscientist at the

University of Queensland who oversaw the study, Dr. Nick Lavidis, says grass is happy-making all on its own. (Yes, we're still talking about grass as in lawns, not cannabis.) When cut, it releases certain chemicals that affect the endocrine system, where stress hormones are regulated. Memory isn't even a part of it. Even if you spent every day of your childhood on city pavements, the fresh-cut grass will still change your body chemistry enough to affect your mood.

All of this brings us back to my neighbor's house-selling gambit. I couldn't find any research showing that fresh-baked bread contains chemicals (like cut grass does) that could improve your mood, so the bread-baking trick works largely by invoking a fond memory. Fair enough—but I don't know very many people who grew up in homes where loaves of bread were regularly whisked from the oven. It's possible that for some people, the smell will zip past the hippocampus without having any effect, and the rational brain will then kick in. *Oh, they're trying to make the place smell like baking. I wonder if that's to hide a smell of mold?* At least, that's what I'd like to think, because it's comforting to believe that we are mostly in control of our thoughts and feelings. But we're not always. A pleasant smell can spin its sweetness onto your mind even when there's no direct connection to your own life, perhaps calling on a deeply embedded evolutionary memory. For our ancient ancestors, the fruits and flowers blossoming in springtime promised safe and abundant food to eat. The pleasurable evocations of sweet smells haven't left us.

Kat Cole, who ran the Cinnabon company in the 2010s, understood the power of our senses to overcome our thinking brain. Anyone walking through a mall or airport in those years remembers the overwhelming aroma of the sweet cakes, as bold as a carnival barker enticing you to come in. Cole, who was then in her early thirties and had already been president of Hooters, knew that a visceral experience can be the best tool for big sales. If you're going to convince people to snack on something that has 880 calories with five times as much fat and sugar as a glazed doughnut, you can't let the rational brain play any part.

Cole went on a sensory offensive. She put the ovens in the front of the stores so the scent could easily escape into public spaces, and she advised store managers to put sheets of cinnamon and brown sugar in the oven every thirty minutes to intensify the aroma of the baking buns. "The lure is something ooey gooey, totally delicious, and incredibly indulgent," she said brightly. To prevent the smells from disappearing up air ducts instead of penetrating the pedestrian areas, she got the weakest oven hoods allowed by law. If you're the type who usually eats yogurt but found yourself drifting into an airport Cinnabon one day, you can congratulate Cole for understanding the power of our senses.

The more we recognize the influence of the physical experiences around us, the more we can understand the effect they are having on our happiness. If you want to eat a Cinnabon, you should certainly go ahead. But if you've been lured to an experience that you will later regret, it helps to recognize how you're being influenced. Used right, our senses can increase our positive feelings. The secret is knowing how to use them to make yourself happy—rather than just stuffed with buns.

Touching the Moon

Getting the body involved in any hobby or activity has an effect on how we feel—and the more physical it is, the more measurable the effects. A mind-centered view of love and happiness would say that you fall in love or feel joy—and your brain releases the hormones that support and amp up the feelings. But as with so many other actions, we are now discovering that the mind isn't necessarily the driver in this game of love. When you hug or have sex or breastfeed a baby or pet a dog, oxytocin gets released. It's sometimes called the "cuddle hormone," because it's believed to create the emotional bonding that comes with physical intimacy. The sensory experience leads to the emotional feeling.

In a study done a few years ago at UCLA, ninety-five people agreed to be part of an experiment looking at how oxytocin is released. Sixty-five of them had their blood drawn and then got a fifteen-minute massage. (Getting a pleasant back massage is one of the better perks of agreeing to be a research subject.) Those in the control group just sat quietly. After the massage, oxytocin levels had gone up in the massage group but not in the others.

Comic actor and former late-night host James Corden once joked that he finds the whole concept of a massage to be very strange. He feels weird telling his wife that "someone is going to rub my entire body and they're just going to miss out a tiny little section." It's a funny line—but he has a point too. The same hormones and neurotransmitters that get involved when you have passionate sex with a lover also come into action during a massage. Ancient Greek physician Hippocrates thought massages made people happier, and he was amusingly accurate in referring to it as "rubbing."

I admit that I'm in the James Corden school of massages. If someone is going to rub my body, I want them to like me first. My friends who get massages find this enormously amusing and point out that the pleasure of a massage is strictly physical. I understand, and most masseuses and their clients know exactly where the line is drawn. But our bodies and brains are tightly connected, and the oxytocin release can sometimes get confusing. I found several Reddit discussions where people mused about falling in love with their massage therapist. A friend of mine in California asked the person he goes to for spa treatments if he'd ever had that experience—and he got a quick nod of the head. "People love your touch, and they start to associate that with other feelings," the professional masseuse said. "It happens the other way too. You're massaging someone, and the brain starts to think that you must care deeply about them."

Touching is a great source of pleasure, and I'm not suggesting that you give up massages. Anything that increases the release of happiness hormones has a value in making us feel more positive, eager, and motivated. Touch may also be one of the most important physical experiences we have for feeling safe and calm. Babies want to be held and stroked when they're agitated, and grown-ups do too. The first place we seek emotional comfort is with our bodies—touch can bring reassurance and closeness and a sense of safety. Therapeutic touch can also be very powerful. As your body relaxes to a professional touch, your emotions become less restrained and tension gets released. Being touched gives you a sense of where your body is in the world and reminds you that you live in that body. It not only lets you move through the world—it holds all your emotions and feelings.

The eighteenth-century German philosopher Johann Gottfried Herder wrote extensively about aesthetics and our senses. In trying to understand what makes something beautiful, he said that sight, considered the "highest" of the senses, wasn't nearly as important as touch, which he thought had the power to shape how we understand the world. "Everything that has form is known only through the sense of touch," he said. He thought that to a blind person, the sense of touch would be a full replacement for the sense of sight. In an inordinately moving passage he cites an essay by the philosopher Diderot and says that if a blind person "wished for an increase of his senses, then it would be for longer arms to be able to feel the moon's surface with greater clarity and certainty, and not for eyes to be able to look upon it."

Longer arms to touch the moon. What an extraordinary idea! And a glorious reminder that there are many different ways we can let our bodies bring us joy. Since everyone seemed to take a shot at rewriting Descartes's "I think therefore I am," Herder came up with his own briefer and very potent version: "I feel! I am!"

What Jellyfish and Cars Have in Common

On a recent trip, I excitedly went down to the dock at our hotel the first day and jumped into the bracing water of the Mediterranean. I felt invigorated swimming through the slightly rough water, and I assumed its iciness explained the tingly feeling in my skin. But soon the tingling felt more like stinging, and I turned back to shore. Once on the beach, I saw my legs and arms had bright red stripes that were beginning to swell. I'd been attacked by jellyfish.

The jellyfish lolling in an ocean or aquarium don't look like much. They're practically transparent with skin so thin it absorbs oxygen directly (no need for lungs). The layer just beneath is filled with nerves that respond to stimuli. The jellyfish is a creature of all nerves and no brain at all. You'd think that my advanced (comparatively) brain would make me the smarter one in the sea—but guess which one of us emerged just fine and which one had whiplash marks that lasted for weeks?

The main goal of any living organism is to stay alive—and

whether you're a jellyfish or human being, your system is first programmed for survival. Humans have been around for about two hundred and fifty thousand years, which makes us pikers compared to species like horseshoe crabs and jellyfish, which have survived in some form for some 500 million years. Jellyfish are nothing but sensation. When the nerves on their tentacles sensed my presence, they lashed out to send me away. It worked. The jellyfish didn't think about the process (remember, it has no brain), and I couldn't take it personally. It was simply a physical reaction, a sensory response.

The human chain is more complicated, of course, but the basic idea of the senses leading the parade of response is similar. Recognizing that our thinking brains are sometimes less in control than we expect can be enormously liberating. It helps explain actions that you might not otherwise like. Pondering the jellyfish response and our human equivalent even helped repair one of the biggest hitches in my marriage.

Here's what happened. My husband and I spend a lot of time traveling together on long highways and dark country roads. He's a good driver and usually the one behind the wheel— but when you're in the car enough, things happen. Every time he slams on the brakes or swerves away from another car, I respond with a very loud and audible gasp. The shocked cry of distress comes out before I can stop it—and believe me, if I could stop it I would, because my husband finds it incredibly annoying every time. We have probably had more arguments in the car than anyplace else.

The night after the jellyfish incident we got in the car, and I told my husband my new insight about body-mind reactions. When he thinks I'm criticizing his driving, it's really just my body responding before my mind can even get involved.

"You think I'm gasping to make a statement about your driving, but it's a completely reflexive action," I explained.

I pointed out that his action—the braking or swerving—is also instinctive, a reflective response to a danger on the road. His instinct has a practical purpose: survival. Mine comes from the same source with somewhat less valuable results. My body senses danger and sends out a surge of adrenaline, which leads to my sharp intake of breath. You'd think the moment could just pass without any repercussions, but it never does. The involuntary gasp sets up a domino effect of additional responses. (This proves again that we're more complicated than jellyfish.) My mind picks up the changed physiological state—higher adrenaline, quicker heartbeat… the gasp!—and sends the emotional signal that I should be anxious or tense or scared. It takes a minute or two for everything to calm down again. On a dark road at 60 miles per hour, that can be a long minute or two.

If our mind and emotion preceded our physical responses, I would never gasp. I would just tell myself to sit quietly and let him drive. But here we are again—feeling comes before thought. I don't know that this will guarantee peace on future car rides, but it provided a new perspective for both of us. The computer brain is not always in charge. Knee-jerk reaction has a negative connotation because it means something we do without thought. But just as with jellyfish, it is part of our daily repertoire as human beings.

The Heart of Happiness

Because I spend a lot of time thinking about gratitude and positive emotions, I started wondering what signals my body might send that would alert my brain to something good going on. It makes sense that we are wired to react instantly to danger—but how about to joy, beauty, happiness, and love? If we could find the joy triggers—the events that make us gasp with pleasure—maybe we could find a way to be instinctively happier.

It turns out that the body-mind connection is as powerful in making you feel good as it is in warning you of danger. The more research I did on the subject, the more I realized that it's crazy to think that our minds can make us happy without our bodies getting involved.

When we talk about senses, we often think of the basic ones—seeing, hearing, tasting, touching, and smelling. But those are just the sensations we are experiencing from *outside* the body. An entire other set of nerve pathways take the sensations that are occurring *inside* our bodies and send that information along to the brain. That rumbling sensation in your tummy? Your body sends the information, and your mind may decide you're hungry, anxious, or dealing with the effects of too many tacos last night. Amazingly, research studies show that people who are more in tune with their body's internal senses are happier and more positive. Being aware of how your heart is beating or the rhythm of your own breathing makes you more comfortable with yourself and your environment.

The romantic notion that positive emotions like love and happiness are lodged in the body goes back to the ancient Greeks and Romans. Aristotle described the heart as the center of feeling, and when Shakespeare noted in one sonnet that "Mine eye and heart are at a mortal war," he wasn't giving the brain any part in his longings at all. The great nineteenth-century poet Christina Rossetti once exulted that "My heart is like a singing bird"—a lovely description of happiness that you immediately understand.

The firm belief that the heart is at the center of emotion lasted until the Renaissance when better understandings of medicine and anatomy arose. In the early 1600s the English physicist William Harvey figured out how the circulatory system worked and that was that. Scientists eventually accepted that the heart had a strictly mechanical role and was not the center of spirit and love, after all.

As far as medical insight goes, William Harvey had his facts exactly right, but the idea of the body being part of a great passion isn't so far off. Aristotle and Rossetti and Shakespeare were more on target than we realize. Emotions involve more than the mind—we feel them in our bodies. We may even experience the physical sensation first and then try to figure out its source. In one international study, people across different cultures and geographies identified specific emotions as being in the same place in their bodies. The researchers found "consistent patterns of bodily sensations" across both basic and complex emotions. And yes, people in all parts of the world feel happiness and love in their chest. Being lighthearted when you're happy isn't just an expression—it's a very real physical experience.

I don't mean to take the romance out of that joyous feeling in your chest as you skip across a park to meet someone you love, but there is probably a very specific physical reason for it. Researchers are still debating the link between physical sensations and emotions, but a system called the somatosensory network is constantly sending information from body to brain, and there is some indication that the vagus nerve, which connects the brain stem to the chest and abdomen, also gets involved in the body-mind emotion loop. You don't just experience emotions in your mind—they get lodged in the deepest recesses of your body. Emotions obviously change how our bodies feel, but since the systems work both ways, you can also use your body to make yourself happier. Taking a deep, slow breath can help relieve tension in your body and lighten the heavy feeling in your chest. As your body feels better, your emotional state subtly changes. You have more control over your happiness than you may realize. You just need to take a deep breath to find out.

The New 4-H Club

Ask scientists now about happiness and joy, and you'll hear a lot about dopamine, serotonin, oxytocin, and endorphins. It's hard to turn those into a great poem, but they are the hormones that our body releases to tell us to feel happy. Since they are the big four hormones involved in happiness, I've started to think of them as the 4H.

I spend a lot of time reading academic studies on happiness, and it's amazing how many of them involve measuring levels of the 4H. For example, it's been shown many times that levels of dopamine go up when we have sex or eat certain foods—which makes sense. We are pleasure-seeking creatures, and our bodies are built so that the things that help the species survive also give us sensual gratification. That way we want to do them again and again. Sex and eating are in that category, but it's a little surprising that many other familiar activities—including knitting and crocheting—also cause dopamine levels to go up. In one study of 3,500 knitters across several countries, more than 80 percent said they felt happier after knitting. I'm all in favor of a nice warm scarf, but it's hard to imagine that, in the great plan of evolution, knitting played much of a role.

Something is definitely going on, though, and it seems to involve once again our brain trying to read what our body is doing. Hobbies like woodworking, cake decorating, and pottery require repetitive movements with our hands, which change the levels of the 4H and have a calming effect on our mind. The body-mind link gets happily activated when you are physically making something using the sense of touch.

Researchers including Michael Norton from the Harvard Business School and Daniel Mochon from Tulane University wondered how people are affected by projects they make themselves. In one experiment, they had some people build a storage box from IKEA while others received one that had been already assembled. They then asked each person to estimate how much they would pay for it. You have to assume that the do-it-yourself boxes weren't quite as perfectly constructed as the premade ones. But over and over again, people valued the project far higher when they had built it themselves, and they were willing to pay significantly more than those who hadn't built it.

In another experiment, the researchers had people make origami frogs and cranes. To an outsider, the finished creations might have looked more like crumpled pieces of paper than ethereal animals, but the creators were delighted with what they had made, and when asked to put a price on them, they thought they were worth a lot. They liked their own origami results even more than those done by experts.

When you make something yourself, you don't just value it more highly, you get a great psychological boost from it. It makes you happy. The involvement of your physical participation in the activity stimulates your mind in a way that gets your whole self more involved. That's true at any age. Legos started out as a kids' toy, but now it's a favorite hobby among high-flying adults who pay hundreds of dollars for complicated building sets. (The 5,900-piece Taj Mahal is definitely not for kids.) Everybody—kids and adults—gets a surge of pleasure from the rhythmic action of putting together the small pieces, and once the project is finished, the builders want to hold on to their Lego creation. It's theirs, it's special, and it makes them happy.

Whether you decide to build Legos or take up an activity like bread baking, weaving, or whittling, you get positive results on two fronts. Our bodies aren't really made to sit still—which is why many people find themselves fidgety when they try meditation or other mindfulness techniques. Carrie Barron, a psychiatrist at the University of Texas Medical School, says a do-something-with-your-hands hobby may be the perfect solution. The repetitive activity of such hobbies strikes the right balance between action and relaxation. Your body is engaged, and your mind can focus on just the task at hand. Once you're done, you quite literally have something to show for it. It's a 360-degree example of body and mind working together to make you happy.

Where to Find the Real You

A maxim dating back centuries warns that *Wherever you go, you take yourself with you.* It's a popular meme on Instagram, and author Neil Gaiman updated the idea for a wide audience when a character in one of his books muses about "people who believe they'll be happy if they go and live somewhere else but who learn it doesn't work that way. Wherever you go, you take yourself with you. If you see what I mean."

I do see, and I think Gaiman is a terrific writer—but the "take yourself with you" sentiment turns out not to be completely true. Who is this real you that you take along?

Maybe you think of the real you as the person lurking inside your brain, making decisions. But your brain sits in the dark recesses of your skull and doesn't do anything independently. It needs information from your body and environment to shed light on any situation. It's certainly true that your brain has processed lots of past experiences and depends on those to make predictions and plans for what's ahead. But those can get overruled by what it's discovering right now. I've always been a big fan of Sherlock Holmes, the fictional detective who makes amazing deductions and uses his brainpower to extraordinary effect. But whether in the books or in the movies starring Robert Downey Jr. or Benedict Cumberbatch (my favorite), the brainy Holmes is regularly dashing about to see things, make observations, and collect evidence. The experiences of his physical world are constantly influencing his mind.

Maybe the real you is your physical self? Not quite, since the skin and bones and slightly graying hair you bring along can change with exercise and diet and a box of L'Oréal. Your emotional state—the "you" who you take along—can also change with the input your body receives through your five senses. It's nice to imagine a core personality that is the real you, formed by all the experiences you've had, but our emotional reactions are directed by the current messages from our environment and the responses of our bodies. You may actually be happier when you take yourself somewhere else because your environment affects your mood, your attitudes, and your sense of happiness in the world.

The sensory input you get sitting on the couch in your living room is completely different from when you are, for example, swimming in the ocean. Your feelings about yourself and the world around you change when you're staring at a brick wall versus gazing serenely across a vast blue sea. If asked in each location how happy you feel, you'll respond differently. You'll also have a different perspective on whether your life in general is positive or negative. My terrific editor Erin took to the road as a digital nomad for the last few years and discovered that working from different locations and environments brought her great joy. Simply knowing she would have a new experience became part of the pleasure, and she noticed her happiness peaking "when I could just step outside my door and be in nature—forests, mountains, lakes, the ocean." It's a very different "you" you're taking with you when you go from sofa to sea.

So where will you be the happiest you? And how can you maintain that state? That's what I set out to discover next.

PART TWO

———

The Best Places on Earth to Be Happy

Where you put your body matters to your happiness and general well-being. Certain settings influence your emotions and neural pathways in profound ways. Here's how you can find the best locations for positive results.

Why Blue and Green Are the Happiest Colors

> Allow nature's peace to flow into you
> as sunshine flows into trees.
>
> —JOHN MUIR

I LIKE GOING ON TRIPS. Who doesn't? You get a break from the dailiness of life and find yourself in a new place filled with fresh sensory inputs. Suddenly you feel more awake and alive. By breaking normal routines, you no longer allow your body to operate on autopilot and even walking down the street feels stimulating. Your body, alert to new sensations, rushes information to your brain, which encodes the experiences differently than on a normal day. Your body sends a message to stay alert—*This is something special! Something we haven't seen before!* The increased sensory awareness is one reason you remember your vacations so well.

A getaway doesn't have to be expensive or fancy to get you stimulated. When our kids were young and our finances meager, we often went on canoe trips, pitching tents at night and cooking over an open fire. To this day, our sons will talk about digging a hole in the ground to bury the leftover chili so the bears wouldn't come. We don't do that at home. Any experience that's a sensory break from your everyday encounters changes the neural pathways and becomes a new part of who you are.

Vacations often involve more time outdoors and in nature than we typically get in our daily lives—and that simple shift has major influences on our happiness. Parents have been sending their children outside to "get some fresh air" for generations—and they were probably right. Outdoor air usually has more oxygen than the recirculated air indoors, and as oxygen levels in your blood go up, your whole body-mind circuit changes. You start to feel happier and less anxious. Your memory becomes sharper and your mind gets clearer. One study out of Cambridge University even found that being outdoors makes children less likely to become nearsighted. (I wish I had known that earlier in life—I've been wearing glasses since second grade.)

Spending time outdoors is even better for your mood if you are in natural environments—green spaces that include trees, mountains, and parks or blue spaces with rivers, lakes, streams, and oceans. George MacKerron, a psychologist at the University of Sussex, has probably collected more data linking place and well-being than anybody else. Instead of trying to track people's general well-being, he wanted to track moment-to-moment experiences. Exactly where are you when you say you're happy? What are you seeing and experiencing?

To find out, he launched the Mappiness project, an app that beeped people at random times and asked them a few questions about their current mood. A GPS system determined their geographical coordinates. At last count, the study had measured sixty-six thousand people, collecting four million pieces of data. Matching mood to place, he concluded that people were happiest when outdoors or in natural environments. "We underestimate how happy being outdoors will make us," he said. After analyzing the data on the moment-to-moment factors that influence people's moods, he concluded that people are "substantially happier outdoors in all green or natural habitat types."

For centuries poets and philosophers have extolled the virtues of being outdoors, and it's nice to have rafts of data confirming that they were right. MacKerron's data shows that being outdoors increases happiness as much as being with friends (instead of alone) or doing an activity you like (versus one you don't).

"We've evolved to find natural spaces restorative and restful. They create happier moments," says MacKerron.

Creating happier moments has a longer-lasting effect than you may realize—because what is life other than a collection of the moments we create? It may be that the more moments you can experience outside, the happier your overall life-view will be. When you close your eyes and think of moments you experienced joy, gratitude, or happiness, it's unlikely that the image involves a windowless setting.

Even hints of nature in a city setting can improve your mood and well-being. MacKerron began his project in the UK, and the pro-outdoors data started showing up immediately. He found that positive moods increased when people were near public parks or trees in the street. Being close to the Thames River or even a narrow canal also boosted mood.

Other research also supports the nature-is-good-for-you idea. In one well-known study patients in a hospital recovered more quickly from surgery and had less pain when their window looked out on a green park or trees rather than a brick wall. The view of a natural setting from a window at work also makes people like their jobs more. In a study I've always admired by psychologist Marc Berman, people did better on memory tests after they took a fifty-minute walk in a country area versus spending the same time walking through a city. Berman has attributed this to what has been dubbed "soft fascination." The sounds and sights you encounter in nature can engage your mind but aren't distracting in the way that the neon lights and big crowds of an urban area might be. When you are engaged by the soft sounds and sights of nature, your body relaxes—there is nothing to put it on high alert—and your mind can wander to other topics and problems.

Interactions with nature come in many forms. Swimming in a lake may give you one set of benefits including a full transformation of your sensory experiences. Walking around the lake or just sitting on the banks looking out results in other physical changes. Berman and several colleagues recently presented a report with recommendations on how to use nature to improve mental health, breaking down the natural experiences by different "doses" and "exposures." The language says it all—suggesting that nature is the best medicine.

Being in natural settings affects your body in many ways. Several studies show that your heart rate and blood pressure go down when you're outdoors, and one meta-study found wide evidence that being in natural environments reduces cortisol levels—commonly considered a biological marker of stress. A dose of nature may be as powerful as a dose of an anti-anxiety drug. Even better is that sunlight, green parks, and flowering trees are almost never toxic. Strange, then, that we are more likely to trust the mood-booster pill created in a lab rather than the natural environment all around us.

Let the Sun Shine In

Given the power of nature to make us feel good, you won't be surprised to hear that there's a connection between sunshine and happiness. When you describe someone as being a "ray of sunshine" you mean they bring warmth and cheer to any gathering. The link between sunshine and good moods is more than metaphoric. While the differences may be subtle or dramatic, most of us are in better spirits on sunny days than dreary ones. The scientific explanation involves the neurotransmitter serotonin, which has been connected to a greater sense of happiness and well-being. Levels of serotonin seem to rise on bright days. You could say the sunshine gives you a sunny disposition.

A study done out of Brigham Young University found a strong correlation between getting enough sunshine and feeling positive. Other weather-related conditions didn't seem to have any effect. But periods of what the researchers called "reduced sun time hours"—also known as winter—led to more depression and mental health distress.

"We tried to take into account cloudy days, rainy days, pollution…but they washed out," says psychologist Mark Beecher, who led the study. "The one thing that was really significant was the amount of time between sunrise and sunset."

Some people respond so strongly to the lack of sunshine that a diagnosis of Seasonal Affective Disorder became popular for a while. The acronym SAD makes for good headlines, and many companies now sell light treatments to mimic the lost sunshine and improve mood. I'd say the jury is still out on how well that works, since it's unlikely that anyone plunges into depression exclusively from the weather. But if sitting under bright lights or traveling to a sunny spot improves your mood, it's worth a try.

We may not have any control over the weather, but it has more control over us than we usually realize. A researcher who studies decision-making analyzed 682 actual university admissions and determined that "clouds make nerds look good." His findings showed an amazing connection between the weather and the way specific achievements were viewed. Specifically, "an applicant's academic attributes are weighted more heavily on cloudier days and non-academic attributes on sunnier days." In other words, the captain of the lacrosse team might look appealing to the admissions officers on sunny days, but students relying on straight A's in calculus and physics to make an impression better hope for some cloud cover when their applications are read. The findings were so statistically significant that he concluded that the degree of sunshine or clouds in the sky on the day an application was reviewed could change your likelihood of being admitted by some 12 percent.

The weather affects you in other impressive ways you never realize. One study that looked at twenty-six stock exchanges internationally found that "sunshine is strongly positively correlated with daily stock returns." When buying a stock, experts say they consider fundamentals like profits, market share, and management potential—but being human, they are also influenced by intangibles that don't hold up to critical analysis. Why would sunshine make markets go up? The researchers speculated that being in an upbeat mood made traders more likely to jump in and buy.

You probably don't want to sell your Microsoft stock based on a weather report, but the bigger story is compelling and has ramifications well beyond the NASDAQ. When you're in a good mood you make more optimistic choices. You become less critical and more likely to act on instinct and impulse.

That may explain why gift shops on Caribbean islands get away with selling stuffed alligators and neon light-up shot glasses—and why the trinkets never seem so amusing when you get them home.

Sunshine increases our serotonin levels, which makes us happier and changes our behavior. Our brains feel the positive charge but are sometimes slow to decode the source. People buying stocks on sunny days will insist that they like the stock's economic prospects and would no doubt scoff at the idea that a weather-related good mood played a role in their decision. What's going on? Their brain reads the positive signals from the happy neurotransmitters that are surging with the sunshine—but gets the story wrong. Instead of *Beautiful day making me happy!* it reads *Great stock getting me excited!* Once again, our old friend misattribution is doing its devilish work.

I thought about all this the other morning when I woke up to a rainy day and noticed I had an appointment at the DMV on my schedule. I'd been there the week before to change my license to a Real ID, and a grumpy agent had (for no reason) rejected one of my identification forms. I had new forms that should be ironclad, but just as I started out the door, I had a thought. If a gray day can make us all more churlish, what would it do to the already cranky clerk? I quickly went online and changed the appointment. I can't say for sure that the sunshine the next day helped things go smoothly—but I smiled more and so did the clerk. He thought my forms looked just fine.

Why Tree-Huggers Got It Right

The power of nature to make us happy has been invoked by everyone from Henry David Thoreau's transcendentalists in the 1800s to the tree-hugging hippies of the 1960s. They were correct—and even those tree-huggers who wanted peace and love were going in the right direction. Hugging a tree—or even just brushing up against nature—encourages the release of oxytocin in the body, the same hormone that has been linked to feelings of love and connectedness. In the spring of 2020, forest rangers in a part of Iceland began clearing paths to trees so people could get close to them. If you couldn't hug a friend in those early days of the covid pandemic, you could hug a tree. The manager of one of the National Forests explained that the feeling of hugging a tree starts in your toes and goes up your body and into your brain. It gets you ready to face a new day and new challenges.

Part of the pleasure of hugging a tree is that it (more or less) hugs you back. German naturalist Peter Wohlleben, who wrote the international bestseller *The Hidden Life of Trees*, has shown that trees are both sociable and sensuous, keeping each other alive and finding ways to communicate. Maybe we can't always hear it, but we can feel it. Other forest ecologists like Suzanne Simard say that trees form a connected community, redistributing resources like water and carbon and nitrogen underground for the good of all of them. She says that trees respond to each other by emitting chemical signals that are similar to human neurotransmitters.

A study out of the University of Surrey found that hugging trees can lower levels of stress hormones and reduce blood pressure. (If hugging a tree makes you feel silly, that blood pressure might soar right back up.) You can also think of hugging more metaphorically—the idea is to embrace the woods with all your senses. Touch the smooth bark of a tree and breathe in the rich smell of pine. Being in the woods, your body sends a message of tranquility and comfort. Your physiological systems calm down and your mind responds.

While we're celebrating tree-huggers, I should note that their companions, the advocates of flower power, are also on to something. Psychologists have all sorts of reasons for why flowers give us so much pleasure, and there's some evidence that simply looking at flowers causes an increase in the feel-good hormone dopamine. In earlier days, people who relied on nature for sustenance saw flowers and knew that food and berries would quickly follow. Maybe we're still wired to get that sense of hope and expectation when we see flowers and to feel gratitude for Nature's bounty.

High-end florist Lewis Miller has done public installations—what he calls "flower flash"—in front of subways and over construction equipment. He filled a trash can one night near a hospital in New York City so it was bursting with huge swaths of flowers and cherry blossoms. It was his way of expressing gratitude to health care workers. His goal is always to make people happy. As he put it, "Flowers are mood elevators and stress relievers."

It's not a surprise that a florist thinks flowers are a good way to give and get gratitude, but research bears him out. Working out of Rutgers University, the psychologist Jeannette Haviland Jones did several studies on the effect of flowers on mood. She found that flowers have an immediate impact on happiness and continue to boost your spirits for days. In one study she sent three different gifts to people—a candle, a fruit basket, or flowers. The people who got the flowers reported being less depressed and less anxious than the others and reported higher levels of gratitude. She concluded that flowers have "strong positive effects on our emotional well-being."

For many years when our children were young, my husband brought me flowers every week. No matter what I was doing, I would stop to trim the stems, arrange them in a vase, and give him a kiss of thanks. Bringing flowers can sound like an overused custom, but it just makes us feel good. Anything that improves your mood and well-being can never be overused.

Scientists continue to try to figure out what advantage some flowers gain from their gorgeous colors and exquisite blooms. Whatever it may be, we gain a lot just by looking. When you have beautiful flowers in front of you, you don't have to think about how grateful you are. You just feel it.

At our country house recently, I picked some white flowers that were growing at the edge of our woods. Bringing them inside, I put them in a vase and realized that having them nearby in my kitchen really did boost my mood. When my husband came home, he pointed out that I had picked weeds, not flowers. Okay, I don't have much of a green thumb, but they still made me happy. Sometimes the difference between weeds and flowers is just your own attitude.

The Joy of Blue

On a book tour stop in Chicago a few years ago, I had a busy schedule of talks and media appearances planned. I had visited Chicago before on business, and I expected this trip to be similarly work-oriented and stressful. But after taping one radio show, the host asked if I'd like to join her for a walk on the Lakefront Trail. Half an hour later, we ambled along a path with a lush park on one side and the sparkling waters of Lake Michigan on the other. Kayakers waved to us and a group of paddleboarders glided by.

"This is beautiful!" I said.

She nodded in agreement. "Being by the water makes me love Chicago," she said.

At her encouragement, I woke up early the next morning to check out the other waterfront area in Chicago—the mile or so Riverwalk that slices through the center of the city. Even though it was early, lots of people were strolling or bicycling, and I stopped to sip a cappuccino at a café next to the water. I felt relaxed—being at the riverfront made the day feel more like vacation than work. I headed off to my next interview feeling unexpectedly calm and happy.

I'd already discovered so many ways that your body can make you happy (or not), and now I had my own confirmation *that where you put your body* has a profound effect on your mood. Being near the water in Chicago had been surprisingly joy-inducing—but I shouldn't have been surprised. Research shows that people have more pleasurable moments when they're near blue spaces—rivers, lakes, canals, streams, and ponds. Being near water versus being on a city street is about the same difference in happiness as doing household chores versus going out socializing with friends.

Water seems to weave its magic in a variety of ways. More than any other location in nature, it is constantly changing. At the ocean, waves come in and out in constant motion that is both regular and ever-changing. On a lake or river or brook, the water is always moving, its surface changing with the air currents and its color reflecting the sky and sun. It is the ultimate in the soft fascination of nature, offering a view that is interesting enough to engage our senses but not so overwhelming as to distract from wandering thoughts. The simple rhythms of water and nature induce a sense of calm that registers in your body. Clenched muscles may relax, and levels of the stress hormone cortisol drop. Blood pressure decreases, and a fast-beating heart gets soothed back to a normal pace. Our brains are quick to pick up the physical changes and translate the positive body vibes to a greater sense of overall well-being.

To understand more about the power of blue spaces, I

got in touch with psychologist Mathew White, who has done some of the most impressive work anywhere on environment and well-being. Among other projects, he led a research study across eighteen European countries investigating the health benefits of being near natural places. White himself used to live by a beach in the UK, but after Brexit, he moved to Vienna and lives on the edge of the (formerly imperial) woods. When I told him what I was thinking about water and emotion, we immediately clicked. Knowing my interest in gratitude, White said that when you ask people "What were you grateful for today?" the answer often involves being outside. "I picked up on that, and I wondered—why is nature so good?" he said.

Curiously, many of the studies that have been done on nature and well-being excluded water from the equation— "They thought it would confound the results because everyone loves water. The positive effects were just too big," White said with a laugh.

I asked White if he thought blue spaces had a different impact than other scenes in nature, and he quickly explained that no rivalry existed. "It's not blue versus green. The most appealing places are actually at the interface—like being on a riverbank or coastline. People aren't necessarily happy being stuck in the middle of the ocean on a boat."

Water has immediate effects on us, creating visceral experiences that we register with our bodies. The gentle sound of a babbling brook, the coolness of a rushing stream, and the glimmers of sunlight on a lake manage to be both soothing and engaging. The reflectiveness, color, and sound of water capture our attention and stimulate emotion. Not everything about water is wonderful, though. "If you have a twenty-foot wave bearing down on your house, that's probably not so restorative," said White. Having done studies in Southeast Asia, he doesn't minimize the dangers of tsunamis and dengue fever, and he knows the hazards of water can scare people away from seeking its virtues. "We have to be honest about the pros and cons, but up to now, the benefits to mental health haven't been properly quantified. There's a balance we need to restore. That's my pitch," he added with a smile.

The Call of the Sea Is Seductive

Feeling serene near water may be deep in our evolutionary bones. White points out that when people migrated across the world thousands of years ago, they traveled by waterways and rivers and eventually built their cities near them. "We feel comfortable by water because it's a place we spent a lot of our evolutionary history. It provides certain resources and fulfills some physical needs." In Greek mythology, the world was divided into three parts—with Zeus given the sky, Hades the underworld, and Poseidon the sea. Ancient Greek, Roman, and Egyptian cultures (among others) had many water gods with primordial powers. "There's something deeply inherent in our cultural dynamic that sees water as one of the big symbolic natural phenomena that we need to pay attention to."

In one of my favorite books, *The Awakening* by Kate Chopin, the heroine is drawn to the ocean in times of confusion and distress. The author captures the rhythm and power of the water to soothe: "The voice of the sea is seductive, never ceasing, whispering, clamoring, murmuring, inviting the soul to wander in abysses of solitude; to lose itself in mazes of inward contemplation." The story ends sadly, but even as the heroine gives in to a final despair, it's amid the welcoming murmurings of the sea.

Some of the most romantic cities in Europe have waterways as part of their charm. The bridges crossing the Seine in Paris feature in endless movie love scenes, probably topped only by the canals that traverse Venice. In London, it's a joy to walk along the Thames, and the picturesque canals in Amsterdam are a hallmark of the city (and far more appealing than the red-light district). If you happen to visit the Netherlands on King's Day, the national holiday celebrating the royal birth (really just an excuse for an annual party), you'll find the canals jammed with boats full of revelers cruising through the city.

I could go on and on about great waterways around the world and never run out of examples since some 71 percent of the earth's surface is covered by water. The U.S. has many dramatically blue spaces. In Michigan, 41 percent of the total area is occupied by water, and nobody lives farther than six miles from a body of water. If you live in some of the drier western states in America, that may sound surprising, but many cities around the country, including Austin, Boston,

Seattle, Minneapolis, and San Francisco, have spent billions of dollars reclaiming and renovating their waterfronts into public spaces. New York City recently spent $100 million to build a pedestrian walkway that runs three-tenths of a mile along the East River. Whether an eight-block esplanade is really worth that much money is subject to debate, but I understand the impetus. It's intended to connect to other trails by the water that the city has been developing for years and that have been transformative to their neighborhoods. People feel good when they are near water.

When cities spend money reclaiming local waterways, the expense pays off with better mental health and happiness. As White puts it, "The social return on investment is very high." In one case, his team made improvements to a beach in a working-class neighborhood in the UK that's intriguing to Americans because it's where the *Mayflower* launched. The historic site had become part of a neglected, rundown, and garbage-strewn region. White worked with a landscape architect to improve access to the beach and also create a theater-like space where people could gather and simply view and enjoy the sea. He collected data on people's moods and happiness before and after the renovation. "Health, well-being, and life satisfaction were all higher afterward," he said.

People liked being around the water. Young people had always come by to hang out and drink beer ("I was a youth once so I'm not going to knock it," said White) but now the area became an increasingly heterogeneous community space.

Older people started having lunch together, and visits from mixed age groups went up at all times of the day. "I expect if the view had been a train station or even a normal park, the effects wouldn't be the same," says White. "The water becomes a catalyst for people to gather."

When the actor Rainn Wilson did a TV series on the places people are happiest, his first stop was Iceland. Though the country isn't as frozen as its name would suggest, it's not exactly balmy either—averaging about thirty-three degrees Fahrenheit in winter and fifty-five degrees in summer. But Iceland is known for its hot springs, which are a center of activity and community in most towns and cities. The ritual and tradition of gathering around these geothermal spots may well be a key to the country's high level of bliss.

"Icelanders are a very social people," Wilson says. "They love to sit together in bubbly, hot springs."

In America and elsewhere, people pay a big premium to live near water. Lakefront houses often cost 50 percent more than those a block or two away, and if you want to be near the ocean, expect to fork over millions. But White found that you don't have to live within shouting distances of the water to get the benefits—just making recreational visits has dramatic positive effects. People who have contact with blue spaces for just 120 minutes a week report better health and a more positive sense of well-being than those who don't have the same experience.

"Whether you get the time as a two-hour walk on a Sunday

or a half-hour several times a week doesn't matter. However you can weave it into your week will improve your well-being," says White.

The impact of water on our emotions is so potent that the waves don't have to be lapping at your feet to make a difference. In one study, women pedaling on an exercise bike reported being happier looking at a video screen with a seascape (blue group) or a countryside (green) rather than an urban setting or a blank wall. Those in the blue group thought the time on the bike had gone by the quickest and were also the most likely to want to do the exercise again.

Even something as simple as a fountain gurgling in a city square can be surprisingly enticing. Water makes us feel hopeful and lucky—which could explain why we toss coins into fountains then look with satisfaction at them resting on the bottom. What would Rome be without its Trevi Fountain? So many people are beguiled by its flowing streams that more than a million and a half dollars' worth of coins get tossed into the fountain each year. (They are eventually collected and donated to local charities.)

After reading all the research on blue spaces, I ordered a mini-waterfall for a table in my living room. It cost $35 and had a small motor that sent water spilling over some gray rocks. It was neither a huge investment nor particularly chic, but I set it up one winter evening when we had several friends coming over for dinner—and everyone commented on it. At various times, I noticed people getting distracted from our conversations and just staring at the spouting water. Maybe if it hadn't been dark, they would have been more entranced by the view out the window, but this way, there was a bit of nature inside. The calming gurgling sound and the mesmerizing flow of the water changed the temperament of the evening.

I asked Mathew White if he thought a glowing fire in a fireplace could play the same role that water does—giving us something quiet and ever-changing to look at, a background that gives comfort and diversion without being too distracting.

"Absolutely!" he said. "When you're gazing at a fireplace or watching water, you free your mind from immediate attentional needs and allow it to wander. Where does it go? I'd love to find out. Hopefully, someplace happy."

The Magic of Rainbows

The great naturalist E.O. Wilson suggested some forty years ago that people have an innate desire to associate with other living things. The "biophilia" hypothesis has since been used to explain everything from the reasons we bond with our pets to the positive impact of having growing plants (as opposed to plastic ones) in your office. It's the reasoning behind the increased acceptance of emotional support animals (you can take a pig on a plane if it helps your mental health) and why one man I know was allowed to keep a dangerously large tree growing on the patio of his tenth-floor rental apartment when he said it eased his depression. Try that with a pile of bricks (or even a pile of books) and you won't get very far. Many researchers seem smitten with the idea that people seek connections with other vital life forces—and the biophilia concept gets bandied about to explain the many positive emotions we experience in natural settings.

But something else must be going on. Biophilia is about biological connections, but a whole range of nature exists outside of that realm. White notes with some relish that our much-beloved water is abiotic—meaning it is not living and so doesn't fit into the biophilic theories. ("There may be a few fish in water, but we're not talking about fish," White said.) Like rocks and rainbows, water is an environmental phenomenon that is not alive but draws us in anyway. If a biological connection isn't the answer, what is the magic key?

My house in Connecticut has a westerly view, and on summer evenings when the setting sun sends off brilliant streaks of orange and red, everyone rushes to the deck to look. Whoever has been in the house—kids and grown-ups—stands watching the light show and feeling bursts of joy. What is it about sunsets and rainbows and other natural sights that makes us stop and look and sometimes gasp in delight? What are our bodies experiencing at those moments that makes us so happy?

"That's the million-dollar question," said White.

The answer may be in what White describes as the importance of ephemeral phenomena. With any change in the environment, our physical sensitivities increase and we become more alert. That tingling of excitement you feel is very real as every part of your body prepares to experience something new.

Part of the enchantment of the sunset is how quickly the sky bursts into glorious color and then vanishes into darkness. You want to experience the fleeting moment before it disappears. Similarly, when you see the glowing colors and distinctive arc of a rainbow, you stop immediately to take a photo because you know that in a moment it will be gone.

"If a rainbow were in the sky twenty-four hours a day, seven days a week, it would just become part of the background," says White. "But what's tantalizing is that you need to appreciate it right now."

The burst of delight you experience on seeing a rainbow or a sunset—both of them unexpected and ephemeral—is very different from the pleasures that come in the green spaces of nature. Amid trees and mountains, change occurs slowly, lulling us into comfort. The soft fascination of green spaces allows them to become a background to our thoughts. Ephemeral phenomena capture us immediately. We pay full attention.

"It's a bit like the magic under the mistletoe," says White. "Enjoy it now because it will be gone in a moment."

Some longer-lasting events and places can also make us happy. How do you know what they are? I discovered next that it's not always what you expect.

6

Places That Make Your Spirits Soar

We shape our buildings; thereafter they shape us.
—WINSTON CHURCHILL

Listen to silence. It has so much to say.
—RUMI

EARLY IN MY MARRIAGE, I stood in front of a tiny cottage by the sea and announced to my husband that "I could be so happy" if we lived there. The cottage wasn't for sale, and we couldn't have afforded it anyway, but that started my continuing quest to find the perfect place to be happy. I have since said that "I could be so happy here" at a farmhouse in Vermont that looked out on a field of grazing Angus cows and in a town called Soglio that we had walked for five days across the Swiss Alps to reach. At the time, it had no internet, minimal cell service, and nobody spoke English. Maybe I could be happy looking at the view, but I wouldn't be able to do much else.

The place that will make you happy is often the place where you feel most free to let your spirits and imagination soar. Gather any group of writers and artists together and sooner or later someone will ask *Where do you work best?* It's not just a question of where to find a cheap rent or a quiet space. Your physical engagement with the world becomes an emotional and creative connection too. What we look out at during the day changes how our minds work, how happy we are, and what we create. Looking out on a beautiful sunset or an expansive ocean, you naturally feel a sense of gratitude for the world. You don't have to think about it—the joy comes to you.

A sense of place pervades many great works of art. French Impressionist painter Claude Monet found inspiration at his country home in Giverny, and his famous water lilies paintings depict the scenes he saw in his garden. Giverny now attracts throngs of tourists who visit Monet's house and gardens. Sure, they want to see the beautiful landscapes, but they also hope to drink in a little of the Monet aura. A person comes to define a place and the place defines the person.

Another Impressionist painter, Paul Cézanne, found his perfect place on the other side of the French countryside in a town called Aix-en-Provence. It too attracts art fans who want to stand where he stood and see what he saw. He created some ninety works looking out from a particular hillside called Mont Sainte Victoire, and the town has placed reproductions of a few of them on the mountainside. A studio in Aix-en-Provence where he worked for just a few years (1902–1906) now attracts busloads of visitors and even has a gift shop. "It is here, on the Lauves' hill amid his dearest familiar objects...that you will feel most intensely the full presence of the painter," the website promises.

Looking for the spirit of a great artist in the place he lived or worked is not quite as fatuous as it might sound. Cézanne and Monet and countless other artists looked to the physical worlds around them to ignite their creative brains. Whatever was in the air, the light, or the views played on their neurons and changed the person (and artist) they became. You don't have to believe in ghosts to hope that feeling the same physical inspirations will give you a sense of who they were.

If I give you a choice between going to a beautiful, airy, and sunny location with vast mountain views and spreading palm trees or a dark, dank spot with rusting pipes all around, you'll no doubt go for the first one. The intuitive sense that our well-being improves in beautiful settings seems to be true. Researchers at the University of Warwick did a complicated study where they connected the "scenicness" of various geographic spots around the UK to people's well-being and health. They wanted to go beyond what we already know about green spaces being good for us, and their results showed that in urban, rural, and suburban areas "the aesthetics of the environment may have quantifiable consequences for our well-being." A beautiful place makes you happier, often on a physical level that you don't even recognize.

The World Happiness Report comes out every year with a list of the happiest places in the world, and the winner for six straight years has been Finland. Also regularly hitting the top ten are Denmark, Iceland, the Netherlands, and Norway. The many social, political, and economic factors that go into making these happy countries aren't easily copied by others, but the Finns recently decided to tout the happiness habits they could share. They invited people around the world to apply for a free Masterclass in Happiness, and some one hundred and fifty thousand people signed up for the lottery. The fourteen winners got a trip to a lake resort surrounded by pine forests to learn the Finnish secrets of happiness based around four categories—nature, health, design, and food.

Since all the categories seemed based on the physical experiences of your body, I emailed one of the execs at the Finnish tourist board to find out more details about their secrets for happiness. I got an automatic reply that she was on summer holiday for the next month with no access to email—leading me to guess that one of the secrets is taking long vacations. A publicist for the campaign explained that Finnish happiness is all about a closeness to nature—"our energizing forests, charming lakes, and vibrant archipelagos." Marketing talk or not, she's right that it's easier to feel happy when you are surrounded by the bounties of nature and can take time to indulge in the pleasures of healthy food eaten next to a beautiful lake. It's even better when someone else is paying for it all. Are fragrant pine forests and refreshing lakes sufficient to make all of us achieve a Finnish level of happiness? Probably not. But it may be a start.

The Mathematics of Clutter

Most of us won't move to Finland or Aix-en-Provence or a green field in the middle of nowhere, but we can find closer-to-home tips for creating happy places. Simply making your own space more pleasing to your personal sensibility can improve your sense of well-being. One reason that books, TV shows, and social media posts on decluttering remain so popular is that our bodies feel soothed by clean lines and predictable patterns. A messy pile of dirty clothes or a table with unwashed dishes is jarring. Your body tenses in the same way that it would if a snake were to appear on your hike—*Something is out of place here!* With the snake, your body's response is quickly handled by your brain, which tells you what to do to avoid danger. With the clutter, your body and brain get out of sync. Your body feels something wrong, but since your brain knows that you left the dishes there, it doesn't demand any action. You are just left feeling slightly—off-kilter. Your mood teeters downward.

Startlingly, there's one condition where our bodies become significantly less stressed and our brains produce feel-good alpha waves—and it's when we're looking at fractals. I've heard about fractals for years, and I have to admit I've always found them somewhere between mystical and impossibly confusing.

The basic explanation of fractals is that they are non-regular shapes that look the same at every scale and no matter how big or small they may be. You can't reduce them to anything smaller. The bubbly foam from waves creates fractal patterns and so do the branching streams of a river. Or the spiral shape of a pine cone. The most common fractal we see every day occurs in trees—where the trunk and main branches form an angle that gets repeated in smaller branches, and then the veins on the leaves often have exactly the same pattern.

The word was first coined by the brilliant French American polymath Benoit Mandelbrot in 1975 to explain how things that are normally considered geometrically chaotic—like a shoreline or mountain—can be shown to have a geometric order. He thought these geometric shapes made more sense than the simpler ones we usually consider. "Clouds are not spheres, mountains are not cones, coastlines are not circles, and bark is not smooth, nor does lightning travel in a straight line," he explained in his famous book *The Fractal Geometry of Nature*.

All this would be a nice game for mathematicians to play and for the rest of us to ignore—except fractals may play a role in our happiness. Richard Taylor, a physicist at the University of Oregon, has done experiments showing that looking at fractals engages the same parts of the brain that are engaged when you listen to music. When viewing fractal images, people felt happier and fell into a more relaxed state. "Your visual system is in some way hardwired to understand fractals," he said.

The natural environment is full of fractal patterns—and our bodies feel happiest and most comfortable in environments that match them. The scenes seem familiar and make us feel safe. Go outside with a child sometime when the sky is filled with white puffy clouds and ask them what they see. You'll probably hear about birds and cats and maybe a dragon. It's more than fantasy. Nature's fractal patterns get repeated over and over again, and we take great joy in seeing them.

Cézanne and Monet did not think about fractals (which hadn't yet been named) when they created their beautiful paintings and naturalistic scenes. But they knew that something about the environment made them feel relaxed and creative. Let your body lead you to places like that, and you'll increase your sense of happiness and joy.

The Happiest Buildings

As a college undergraduate, I was one of hundreds of students in a lecture class taught by Vincent Scully, the brilliant art and architecture professor. I'd grown up in the suburbs and knew nothing of the important buildings and modern design styles he hailed. I listened enthralled as he described the great glories of the Seagram Building in New York City, which he said was in the vanguard of a style that had forever changed the world. Scully spoke with endlessly convincing passion, and I couldn't wait to see the building he lionized for myself.

When I took a train into Manhattan for the first time a few months later, I excitedly walked up Park Avenue, waiting to be thrilled. But I somehow walked right by the Seagram Building without ever noticing it. I walked back and forth on the same blocks a couple of times. Wait, was that it? The functional glass building with a plaza out front was—fine. Looking at it, I felt coldly uninspired. Why couldn't I burn with the same passion that Scully felt?

Our built environments sometimes make theoretical sense—at least to the architects who create them—but don't necessarily make our bodies respond with pleasure. The Seagram Building is an example of the International Style that dominated architecture in the midtwentieth century—very flat and geometric buildings often with glass curtain walls. For architects at the time, the minimalist construction and bare materials were a reaction against the fussier more ornate buildings of earlier generations. The French-Swiss designer and urban planner Le Corbusier led a similar movement using rougher materials that became known as the "brutalist" school. You probably have a library, school, apartment building, or city hall in your town that reflects the style. It's functional and utilitarian—but also cold and soulless. The name fits the feeling. Nobody steps up to a brutalist building and feels joy.

If you want to find buildings that make you happy, the curvier lines and eclectic styles of postmodern buildings are a better possibility. One of the leaders of the movement, the architect and theorist Robert Venturi, seemed to get it right when he recommended elements that are "hybrid rather than pure...distorted rather than straightforward, ambiguous rather than articulated...redundant rather than simple." The fractal researchers would probably cheer this view. Taylor and others have found that the skylines that make us the happiest are very different from the "Euclidean man-made shapes such as circles and squares." The gentle curves and intricate patterns that remind us of nature are fortunately returning to architecture in many cities—and done right, they give you an instant sense of gratitude for the richness of life.

Venturi and others are often credited with creating a brilliantly modern concept of design, but architecture that hits you on a visceral level and changes how you feel has been around for millennia. If you're ever traveling in Europe, take some time to visit an old church in almost any small town. It doesn't even have to be one of the famous ones, because there's a good chance that the soaring ceilings and elaborate decorations will immediately make you feel grateful and inspired. When it comes to the great cathedrals, you don't have to be either Christian or religious to get the sense of wonder they create. The wise builders of these structures wanted to inspire a sense of awe when you entered, and they require no theorizing. I've visited cathedrals with people far more knowledgeable than me who explain the details of why the structure is Gothic, Baroque, or Renaissance and what some of the symbols mean. It's all very interesting but somehow beside the point. A great building speaks to your deep-rooted feelings, making your heart soar and your spirit happy. You feel the magic when you step inside.

When I visited Barcelona a few years ago, I went (as every tourist does) to the Sagrada Familia, the great Roman Catholic landmark that was begun in 1882 and designed by the remarkable Catalan architect Antoni Gaudí. You can look up photos of the incredible structure and you'll usually see cranes looming in the background because it's still not complete. Last I heard, it was estimated to be 70 percent finished, and when it's done it will be the tallest church in the world. But even now, some three million visitors a year enter the great church and stare awestruck at the extraordinary spires, sculptures, facades, and designs. I'd never suggest that there's just one thing that leaves you awed and amazed as you wander through (or better yet, as you sit in a pew to avoid the crowds), but it's certainly worth noting that Gaudi is not one for geometric lines. The church inside and out is filled with organic shapes and forms and themes of nature. Your mind may be busy trying to interpret the symbolism—but your body already feels at home.

In the visual aesthetics of what makes people happy, some of the key elements include a place that gives you a sense of openness, a vast scale, and a feeling of naturalness. You experience those pleasures at the top of a mountain (open, vast, natural) and you feel them also in a soaring Gothic church or the Gaudí masterpiece. In those environments, your body tells you that all is well.

In your own house, you can amplify some of the happiness elements by adding eye-pleasing plants like ferns (a classic example of a fractal) or bookends made of natural crystals (also fractal-full). If your house has high ceilings and natural light, you're a step ahead. If you're in a boxier apartment, try to break some of the geometric lines with soft pillows and art or wall-hangings with eye-catching patterns. Designers try to tell us every year what's in style and what's not—but those medieval churches are a reminder that the forms that make us happy never go out of style.

Shhh...Please

When your eyes take in a beautiful scene and you feel a rush of pleasure, you may not be fully aware of what else your body is registering. The sounds of rustling leaves, the whir of a breeze, or the chime of bird songs in the distance all provide a visceral sense of calm even if you don't consciously focus on them. What you definitely don't experience (if you're still happy) is the overwhelming noise of jackhammers or car alarms, or the annoying whine of lawnmowers and air conditioners.

Our body's instinctive response to sound makes loud noise a prime detractor of happiness. When a sound reaches you, it travels through the eardrum and vibrates the bones of the middle ear. In an incredibly complex process that involves the fluid-filled cochlea of the inner ear and thousands of sensory cells, the sound wave then gets converted to an electrical impulse that travels to the brain. You don't have any control over this—your body is on its own.

Your body registers a loud noise as a reason to go on high alert. Mathias Basner, a doctor, researcher, and sleep expert currently at the University of Pennsylvania, has done studies around the world on the effects of noise on sleep, health, and well-being. He says that noise causes the body to release stress hormones like adrenaline and cortisol. Surrounded by noise, you feel more anxious and tense—and less happy—even if you're not completely sure why.

I don't know if the world is getting noisier—but restaurants definitely are. Restaurant owners may think that a high decibel level connotes an energetic and fun place to be, but some combination of loud voices, pumped-up music, and restaurant design has moved the level at many places from fun to painful. In the days when fancy restaurants had cushy banquettes, thick tablecloths, and wallpaper or curtains, the fabrics absorbed some of the din. But now with a minimalist style that includes lots of hard surfaces, exposed stainless steel, and wooden floors, the noise reverberates.

When my husband and I came home from a restaurant dinner with friends the other night, I realized that I felt incredibly cranky. I like our friends and the food had been good, but the room was so loud that we had to strain to hear each other. I felt like I'd spent two hours in a state of tension rather than relaxation.

A raucous room can put you in a lousy mood, but that's just the start. Beyond wrecking an evening, too much noise increases your risk for high blood pressure, heart attack, and stroke. It may have connections to cancer, diabetes, and obesity. A study from the World Health Organization a few years ago looked at environmental threats to our health and put air pollution at number one and traffic noise at number two. They said that in Europe alone, a million healthy years of life are lost every year to the damaging effects of noise.

The idea that people simply adapt to noise has been proven to be not true. You may not consciously hear the disturbance after a while, but your body never fully adapts. Environmental psychologist Arline Bronzaft once studied children in a city school that sat right next to a train track where the very loud trains rumbled by every four or five minutes. She found that by the sixth grade, children who faced the train noise were nearly a year behind in reading scores compared to children on the quiet side of the school. *A year of learning lost to too much noise!* The results were stunning enough to raise a public furor. The Board of Education agreed to install acoustical tiles in the ceiling, and the Transit Authority came up with a way to make the tracks quieter. Bronzaft, a professor at Lehman College, was suddenly a hero. She followed up sometime later and found that once the classrooms were quieter, children on both sides of the school could read at the same level. Bronzaft left academia to work full-time on the problem of how noise impacts communities and diminishes our well-being.

Aside from sticking your fingers in your ears, individuals don't have any built-in way to lessen the noise around us. We have eyelids but not earlids. Even when you're asleep, the eardrum, cochlea, and auditory nerves never shut down. They keep vibrating and sending signals. As Dr. Basner describes it, the auditory system has a watchman function, constantly monitoring the environment even while we're sleeping.

If your bedroom is in a noisy place, your body could be sending out those stress hormones all night, even when you think you're in a peaceful slumber. In various studies that Dr. Basner and his colleagues have done, people subjected to traffic noise woke up often during the night and had severely fragmented sleep. Interestingly, they often didn't remember the disturbances and told researchers that they'd fallen asleep quickly and never woken up. "Great night!" they reported—but the physiological signals that had been recorded showed a very different story. Maybe the noise-induced awakenings were too short to be consciously remembered, but their bodies recorded the problems and the long-term dangers remained.

Noise can upset your happiness during the day too—even when you're not conscious of the problem. As I'm writing now in my favorite library, extremely loud construction is taking place on the street outside. I grumble about how annoying it is, and the friend sitting across from me just shrugs. "I don't even hear it anymore," he says.

I want to tell him that he may *think* he doesn't hear it, but his body is registering every reverberation. He probably has more adrenaline and cortisol circulating than he realizes, and if he goes home later and finds himself short-tempered with a partner, it may be a physical reaction rather than a fully conscious one. I think he should know that. I think it's important we all understand how our bodies are changing us even when we don't completely recognize on a conscious level what's happening. Knowing what's taking place inside the intricate systems of our bodies allows us to be more sensitive to our own happiness, feelings, and well-being. But in this case, I just turn back to my laptop. There's much to be said for the sounds of silence.

People who live in urban areas get used to hearing loud sirens, trucks rumbling on the street, construction sounds next door…the noise assault never ends. In apartment buildings, hearing the footsteps from an upstairs neighbor can be incredibly annoying. Our bodies may be primed to listen for footsteps as a potential threat and so release a flood of danger signals and hormones at those sounds. By the time your brain registers that the only danger is the person upstairs wearing high heels on a tile floor, your whole equilibrium has been disturbed.

White noise machines have become increasingly popular. They work by producing all the frequencies audible to the human ear—supposedly creating a calming sound that masks other noises and helps you concentrate on work or sleep better. I know a lot of people who use the machines at night or download the apps on their phones to be available anytime, but to me white noise sounds like the annoying static you'd hear while trying to find a station on a radio. You can also get pink noise apps, which have an emphasis on lower frequencies (rainfall is often the example), and brown noise, which is even lower. If you like the sound of a storm hitting a beach, brown noise may be for you.

All this could sound mildly entertaining, but Dr. Basner warns that there could be a flip side to having noise machines on all night. Whether the noise is white, pink, or brown, your inner ear reverberates to the sound and translates it to a nerve signal. Your conscious mind might stop registering it, but the body never shuts down its vigilance. The auditory nerve is constantly activated, all night or all day. We don't know what happens when the auditory system is never allowed time to wind down and regenerate.

A physician named Luciano Bernardi did an interesting study on the physiological effects of music on our bodies and brains. He played different tracks of music to a couple of dozen volunteers to see how different rhythms and sound patterns affected them. The music did cause changes in breathing rate, blood pressure, and heart rate, but the biggest surprise came from something he never expected. The two minutes of silence between the tracks had the most dramatic effect. Those breaks proved to be far more relaxing than any of the so-called relaxing music that he played.

I don't have a great solution for the noise problem. Headphones can help, particularly if you're traveling on an airplane or commuting by train. But they don't do much good at a wedding where the band is so loud you can't hear your partner or a restaurant where you want to talk to friends. Perhaps the best solution is to make sure that you occasionally give your ears a respite from noise. Take a walk away in a local park or read a book in a library nook rather than a noisy coffee shop.

The very best idea may be to find a way to listen to birdsong now and then. In one study, people who listened to the sounds of birds online reported lowered levels of anxiety and depression—while traffic noise increased depression. Another study considered real-life contexts, collecting smartphone data from 1,292 participants. The impressive results showed that "everyday encounters with birdlife were associated with time-lasting improvements in mental well-being." The benefits were apparent in healthy people and those who had previously been diagnosed with depression.

The other day, I went outside to the deck of our house and found my husband sitting with his iPad in his lap, just looking off to the distance.

"Everything okay?" I asked.

"The birds are so noisy," he said with a smile.

I sat down next to him and also listened. We could hear the clear whistles of the robins and the trills of the goldfinches, the jumbled warbles of the wrens and the excited whinnying of the woodpeckers. The sounds of the tiny birds carried far distances, and the cacophony was unexpectedly joyous. If I had been part of the mood survey, I would have reported feeling happier at that moment. We are always being told to stop and smell the roses—but for your health and happiness, it may be an even better idea to listen to the birds.

Why Wine Tastes Better in Paris

A mind that is stretched by a new experience can never go back to its old dimensions.

—OLIVER WENDELL HOLMES

HERE'S A LITTLE EXPERIMENT. CLOSE your eyes and think about a time when you were happy and felt a general sense of bliss and well-being. (Go ahead—I'll wait.) There's a good chance that the image that came to mind involved a specific location—maybe a beach or a mountaintop or a cozy pub. It's hard to recall being generically happy. Your happy spot doesn't have to be someplace exotic or expensive. When I asked one friend to try this experiment, she closed her eyes and immediately began to smile. Her happiest image involved a slightly grubby noodle shop where she and her partner first drank sake together and fell in love. Your brain likes to connect an abstract concept with a reality that you can feel or experience.

Once you identify a location as a spot for happiness, your mind will make the connection every time you go there. It's not that your brain is lazy, but since it functions (at least in part) as a prediction machine, it relies on past experiences to anticipate what will happen next. Use that to your advantage. A location that has evoked happiness in the past is carefully catalogued in your brain. Go there again, and the positive vibes you feel are very real.

I've had many happy experiences on trips to Paris, and simply being there now puts me in an upbeat mood. On a recent visit with my husband, I had a bad allergic reaction that required two trips to a pharmacy and a stop at a local doctor. We climbed the well-worn staircase of a classic old building to get to an office where the French doctor sat behind a large wooden desk covered with file folders on one side and a Yorick-like skull on the other. He himself was aged, wearing a blue blazer with a white cashmere scarf. (Fortunately, all he had to do for me was sign a prescription form.)

"That was so much fun!" I said to my husband later, after I had swallowed the antihistamines and we were discussing the day.

Ron looked at me, completely baffled. How could I describe a half-day wasted on medical excursions as fun? I explained that it had been exciting to have this unexpected adventure: the skull, the scarf—what a memorable escapade! I realize now that if I had been anywhere other than Paris, I wouldn't have been quite so charmed. But my brain associates Paris with pleasure, joy, happiness, and fun. Anything that occurs there gets a different spin. Location can change the gratification you take in everyday experiences.

It's Not What You Drink but Where You Drink It

After our trip, Ron went into our local wine store and told the proprietor that we'd had some wonderful wines in Paris, and he wanted to buy a bottle or two to have at home. They discussed the varietals and vintages, and then the proprietor, a cheerful and knowledgeable guy named Ira, nodded.

"I can sell them to you," he said with a sly smile. "Just don't expect them to taste like they did in Paris!"

It took Ron a moment to understand—and then he laughed. The vineyards could export the vintage grapes but not the romantic cafés, the charming streets, or the moonlit views over the Seine. In Paris, we weren't just tasting the wine; we were drinking in the ambiance. You think you are loving what is in the glass, but your senses are drinking in all the experiences around you. The misattribution of emotion that gets people confused on swaying bridges and roller coasters can also make a wine taste better when you're drinking it at a charming café in one of the most gorgeous and romantic cities in the world. The wine might be just the smallest contributor to your happiness, getting more credit than it strictly deserves.

Ira's instinctive understanding that we taste wine (and other drinks) with all our senses has been supported by stacks of experiments. I had a fascinating conversation with University of Oxford psychologist Charles Spence, who has written more than a thousand papers on what he calls "cross modalities." As the head of a research lab at Oxford University, he studies how the multisensory input from our bodies affects our daily perceptions. While most researchers spend a lifetime focused on one sense, like seeing or hearing, Spence was curious about how the information overlaps. "Our senses are talking to each other all the time, often in ways that we don't introspect on or realize," he told me. You don't just see what you see and smell what you smell—your brain has to process the information for it to mean anything. The processing gets complicated. *Where* the sensory input occurs can completely change how it gets perceived, as evidenced by what Spence calls the Provencal rosé paradox. A wine or food tastes different when you have it on vacation than when you try to recapture the experience at home. "Part of it is the atmospheric cues—the smell of the air, the experience of the surroundings," Spence says. "It's also usually true that we're more relaxed on holiday, and things taste better if we're in a good mood."

You don't have to travel very far to have your location affect your sensory experiences. Spence did a wine tasting where three thousand people each drank a single glass of Rioja under four different conditions that included variations in lighting and sound. Most declared the wine "fruitier" when the lighting in the room was red versus green, and "sweeter" when the background music was mellow versus loud. Wine connoisseurs like to discuss the subtleties of terroir and fermentation, but the experiences of our own bodies influence the taste far more than we realize.

The mind-body interaction is a continuous loop of influence. Our brains make a mental model of what will happen, and "We live in the world of those predictions rather than the world of actual experience," says Spence. That helps explain why wine experts given a bottle with an expensive label will rate it highly, even when the wine inside has been switched for a cheaper one. Researchers at Cal Tech and Stanford showed that people don't just *think* the wine is better; it actually registers differently on their brains. Doing MRI scans, they were able to show that different pleasure centers light up when people think they're sipping a $90 bottle of cabernet sauvignon versus the very same wine that they are told is ten bucks. A part of the brain called the medial orbitofrontal cortex gets excited—making you think the wine tastes better.

We tend to think of each of our senses as being discreet and reliable: what tastes sweet in your house should also taste sweet in my house. Any number of location variables can change what we experience though, and it's true well beyond wine. Charles Spence once set up three rooms for whiskey tastings in London's SoHo district and found that by manipulating the environment, he could change participants' ratings of the whiskey's flavor and taste by 10 percent to 20 percent.

One room had green lights, turf on the floor, croquet hoops, and the summertime sounds of lawnmowers and chirping birds. People tasting the whiskey in that environment gave it high marks for "grassiness." In a room decorated in pinks and red with high tinkling sounds coming from the ceiling, the tasters noted the sweet notes of the whiskey.

"People held the same glass in their hand as they went from room to room but experienced different tastes in each place," says Spence. "When they looked at their scorecards at the end, they realized we weren't fooling them—they were fooling themselves." The atmosphere, colors, sounds, and even the comfort of where you're sitting or standing all influence your perceptions.

When Jack Daniels began selling its Bonded whiskey in 2022, the company created a sound track to enhance the experience of drinking, explaining that the high-pitched piano, low and slow cello, and quick-tempo violin would enhance the sweet, woody, and spice notes in the whiskey. The whiskey got excellent reviews, though being 100 proof might have helped people's experience too.

If you want to have a great dinner party at home, you might spend time on things other than cooking the meal. Chef Charles Michel (who Spence describes as "Instagram famous") once created a salad out of snow peas, mushrooms, sprouts, and other ingredients to mimic a Kandinsky painting that he had seen at the Museum of Modern Art in New York. Intrigued, Spence brought him into the lab to experiment with it—and they found that people rated the artistic salad 29 percent tastier than the same ingredients simply tossed in a bowl or lined up.

"The aesthetics and storytelling of what you eat affect how much you enjoy it," says Spence. He also reports that angular plates make people think cheeses taste sharper, and a dessert served on a round white plate is considered sweeter than one on a square or black plate. I can't tell you why strawberry mousse should taste 10 percent sweeter when it's eaten from a white bowl rather than a black one, but he has reported that finding too.

Everything from the weight of the cutlery to the intensity of the lighting can affect how much you enjoy your meal. Give your guests a heavy spoon and they'll think the meal tastes better than if you serve with flimsy cutlery. Adoni Luis Aduriz, who holds two Michelin stars for his Mugaritz restaurant in Spain, has worked with academics to understand what makes a meal truly pleasurable. His restaurant sits on the border of two Basque towns with large oak trees, and you can reach it only by car. He might labor endlessly over his prix fixe menu, which includes more than a dozen elegantly perfect and original courses, but he knows that your happiness in eating the food goes beyond the plate. "Mugaritz is not only the restaurant but also the road leading up to it, the countryside that you can see from the car," the chef says. "Mugaritz is also its setting."

A generous colleague once took me out for lunch at the Gabriel Kreuther restaurant, which also has two Michelin stars. Since the restaurant is in midtown Manhattan, there are no towering oak trees outside, but as soon as we sat down, the maître d' brought over a silk footstool for my pocketbook (perhaps he expected Chanel rather than LeSportsac), which immediately put me in the mood for something special. We sat in a spacious and beautiful art-filled room with wood-beamed ceilings and hand-painted wall coverings. I admired the stunning ceramic plates, each one crafted for the particular menu item, and the excessively elongated cutlery, which felt amazing to hold and looked hand-forged. What did I eat there? I couldn't tell you. I do remember the cloche being lifted from one dish and releasing a great puff of smoke. My happy memories of that lunch rely almost exclusively on the experience of being in the room rather than the actual taste of the food.

I'm sure there are devotees of high cuisine who remember exactly what they ate at any Michelin-starred restaurant and would perch on a street corner to be served the world's best Osetra caviar and truffle paté. But Spence is convinced that anyone who thinks that eating is *only* about the food is kidding themselves. Our bodies take in messages from all our senses and blend them together to determine whether we think something is good or bad, happy-making or not.

One of the best-known chefs in the UK, Heston Blumenthal, runs three Michelin-starred restaurants, appears on many TV shows and has cooked for royalty. He touts the value of "multisensory cooking," and at his restaurant the Fat Duck, about an hour outside of London, he famously serves one seafood dish with headphones so diners can listen to crashing waves, seagulls, and passing ships while they enjoy halibut, mackerel, baby eels, mussels, and more. "We eat with our eyes, ears, nose, memory, imagination, and our gut," says Blumenthal. He explains that food is "the conversation between our brain and our gut, mediated by our heart."

Blumenthal believes that simply hearing the sounds of the ocean can change your experience of eating a snail, and diners who pay several hundred dollars per person to have the full experience at his restaurant probably agree. Our brains rely on the input from all our bodily senses to determine our happiness in any situation or moment. We may not realize exactly what is contributing to our pleasure. The diners at Fat Duck probably marvel at the amazing food they are tasting, and perhaps only a few are savvy enough to say later—*I heard the most wonderful dinner!*

Your Brain Is a Prediction Machine

Expressions like "Seeing is believing" rose for a reason. Until the advent of cameras, photoshop, and Instagram filters, you didn't have much reason to doubt the information that you could see for yourself. Even now when we know we can be tricked, our emotional response is hard to change.

A friend of mine laughs about the different weather apps she and her husband have on their phones. If there's a 40% chance of rain, her app shows a graphic of raindrops with

"40%" underneath it. Her husband's displays a graphic of clouds mixed with sunshine and says "60%." They understand that the apps are actually showing the same forecast with a different perspective. But that knowledge doesn't help much.

"Mine is the grumpy app and his is the happy one," she says. "We can joke about it—but it still affects how each of us feels."

If you look at a picture of rain, your mind registers the visual input and announces *Take an umbrella!* Sure, your brain can do the math and figure out that there's a 60 percent chance it *won't* rain, but we weren't evolved to see the world statistically. If there was a rustling in the trees ahead that might be a tiger, our ancestors had the instinct to run. That's how they lived to become our ancestors.

Many neuroscientists now describe the brain as a "prediction machine" that takes in sensory information and then makes hypotheses about what's going on. There's a constant back-and-forth between the sensory information coming in through our bodies and the perceptions and predictions the brain makes. Sometimes the loop results in an error message. It's a little bit like an optical illusion where you can know the trick but your mind simply can't register it correctly. Another part of your mind remains in control.

What Happens When You Walk on Air

I recently traveled up 1,300 feet to the top of a new building that bills itself as the tallest skyscraper in midtown Manhattan.

Given how many tall buildings there are in the city with famous observation decks (the Empire State Building has starred in numerous TV shows and films), this brand-new outlook was designed to be a full sensory experience. For my guest, I brought along a sweet five-year-old boy I know, figuring it would be a more exciting experience for his innocent eyes than my more jaded ones.

Little did I know.

We took the elevator to the ninety-third floor and stepped out into a huge room of windows and mirrors that reflected and reflected and reflected. We looked out the windows to see New York City below us, then looked down at the floor to see...well, more reflections of us below. We walked through other rooms filled with balloons and clouds to yet another area that included a circular glass floor projected out over the city. Standing on the transparent ground gave the illusion that nothing but air stood between you and the sidewalk a thousand feet below. I watched as people stepped timorously from the solid floor onto the glass ledge. It was perfectly safe, surrounded by floor-to-ceiling glass windows, but some people hesitated, and most seemed at least briefly uncomfortable.

I took the five-year-old's hand, and as soon as the spot was free, we stepped onto the ledge. I immediately felt a strange sensation. Not woozy, exactly, so much as—discordant. I knew the glass beneath me was solid and not going to give way, but every fiber in my body was screaming *Get off! Not safe!*

The five-year-old had no such reaction. He stood firmly in his bright blue sneakers and gave a big smile to the camera. When I showed him the photos where he seemed to be hovering above the city, he looked pleased but didn't seem to find it terribly surprising.

Tottering over Manhattan probably isn't as disconcerting as standing on a glass ledge four thousand feet above the Grand Canyon. You can have that experience at the west rim of the canyon where a u-shaped cantilevered bridge extends out seventy feet, letting you feel like you are walking on air with nothing but sharp cliffs below. I expect that nobody goes to the Skywalk to admire the impressive feat of engineering that took four years to build. You go to be thrilled—and to feel that strange disorientation where your body and brain are having completely different experiences. Since the Skywalk opened in 2007, millions of people have donned paper booties (to protect the glass) and ventured out onto what was for a while the largest glass ledge of its kind. (The Chinese jumped in with an even bigger structure in 2016, a transparent bridge near Chongping that's five yards longer.)

The builders of the Skywalk promise on the website that it's strong enough to hold seventy fully loaded 747 jet planes, so it's not going to crack under your weight. You're safe. Your sensory systems, however, do not read websites. They respond to deeply instinctual neurological feedback that has evolved over the seven million years that hominids roamed the earth and the two hundred thousand or so years since *homo sapiens* first appeared. However rational you may be when you step onto a glass ledge, you can't really outthink that history (and I am extremely rational).

Tamar Szabo Gendler, a chaired professor of philosophy at Yale as well as a professor of psychology and cognitive science, recently came up with a new concept to explain what's going on in some of these situations where our knowledge (*The glass is safe!*) doesn't connect with how our bodies feel (*This is dangerous!*). She says that in addition to our rational beliefs, we have what she has dubbed *aliefs*—patterns that have been laid down in our minds "as the result of our experiences and those of our genetic ancestors." A belief is intentional and can be changed (not easily, but it does happen). An alief, she says, is "associative, automatic, and arational." It happens without your conscious thought and often *conflicts* with what you know and believe.

Much-admired psychologist Paul Rozin has been looking at this kind of mind-behavior discordance for years. He once gave people a brand-new bedpan, completely clean and sterilized, and asked them to drink soup from it. Their response was exactly what you're feeling right now—*Yuck, no!* He also offered people fudge shaped like dog poop and—well, the very thought is so gross I can barely write about it, never mind eat it. Nobody else ate it either. And let's not even get into the details of the experiments with asking people to put perfectly clean fake rubber vomit close to their faces. People know that the items are harmless and safe, but their bodies have a completely different reaction.

In most body-mind interactions, you have an automatic physical response like a pounding heart or surging adrenaline, and your mind scans the symptoms, context, and environment to figure out what's going on. Eventually mind and body come into agreement. The glass ledge situation is different because your sensory input is fighting with your conscious brain—and there's no way to settle the argument.

Walking up a trail on a leisurely hike, you rely on visual input to decide when to step over rocks and how to stay away from dangerous precipices. The visual input you get makes perfect sense to your conscious mind, so all your actions and reactions proceed smoothly (except if you trip over a root, as I so often do). When you step onto the glass ledge over the Grand Canyon or one thousand feet above the sidewalks of Manhattan, though, your visual system tells you that you're striding off the edge of a cliff. Your conscious brain tries to intervene with the information that it's okay. But this time the *all clear* doesn't have much effect.

All this made enormous sense to me—but I was still puzzled as to why the five-year-old was unperturbed. He was getting the same visual input, but however smart and mature he may be, a five-year-old has very little planning or executive function going on in his young brain. The frontal lobes, where those actions take place, are the last parts of the brain to mature. They're still growing by the time children become teenagers, and some research says they're not fully developed until the third decade of life. A child's "prediction machine" isn't yet running at full speed. It doesn't know how to take in the visual input and translate it. A child's undeveloped ability means that parents and caregivers need to function as his grown-up brain. On the positive side, it makes his body-mind conflicts much less dramatic.

The Baker Baker Paradox

Our body is always sending sensory information, but our brain takes in only the tiniest portion. It has happened to all of us that you head down the street looking for a particular store and then discover that you've walked right past it. When you turn around and get back to it, you wonder how you could possibly have missed the bold sign and bright window display. It's not that our brains are lazy, but they're not looking to process any more information than is strictly necessary. Millions and billions of bits of visual information bombard your eyes every second. When light hits the retina, it sends an electric signal to the brain via an optic nerve going directly to a part of the brain called the thalamus—which is also getting input from other senses in your body. The thalamus is pretty talented at combining and repackaging the information coming in and dispersing it to other parts of the brain, but not everything makes the cut. As you walked by the store the first time, your eyes "saw" it, but the brain just didn't register the event.

Our brains evolved to take in information from the physical environment and help us survive. Generally, what we can see, feel, and touch is much more powerful in affecting us than topics that don't have a physical impact. If you have some difficulty understanding abstract concepts in subjects like theoretical physics and non-Euclidean geometry, you're not alone. Your brain evolved to focus on immediate, pressing concerns—and those came in through our physical senses. The best forward-thinking schools are encouraging children to learn through physical experiences rather than just sitting at a desk. Think toddlers happily singing "Head, Shoulders, Knees, and Toes," learning their body parts by actually touching them. As they get older, building the Golden Gate Bridge out of blocks and string will give students an understanding of geometry and physics far more than staring at a workbook. Neurobiologists talk about the Baker Baker paradox where it's easier to remember that someone is a baker than that his name is Baker. Hearing the profession evokes sensory elements—you can smell the fresh bread and see the white hat. A random name disappears into the ether very quickly because we have no associations for it. The connection and the embedded memory become even more dramatic and long-lasting when you don't just envision the association, you have a physical experience for it. If the person telling you he's a baker is also handing you a chocolate cupcake that you taste appreciatively, he's most likely to be implanted firmly in your memory.

All this good information about the power of our senses has led to some interesting ideas about how to improve your memory. One strategy I have seen involves the power of location—linking items you want to remember to a specific place. To recall a grocery list of oranges, bananas, and Oreo cookies, you can mentally walk through your house and visualize the oranges on the foyer table, the bananas hanging from the chandelier, and the cookies climbing up the staircase. When you get to the grocery store, all you have to do is take that same mental walk-through and you'll see (and buy) everything you need. I admit that I have never in my life been able to do this successfully. Most of the time, our brains need more direct physical information. If I see the empty fruit bowl in my kitchen, I will remember to buy bananas.

My children used to like a picture book called *Don't Forget the Bacon,* where a little boy sets off with a shopping list from his mother, who asks him to buy "six farm eggs, a cake for tea, a pound of pears, and don't forget the bacon." He heads out with his basket, repeating the line, but every time he passes something interesting, his mind makes a substitution. Six farm eggs becomes six clothes pegs. A pound of pears is a flight of stairs. A cake for tea is a cape for me. The rhythmic rhymes and illustrations are charming, and I don't think author Pat Hutchins had any deeper purpose than delighting her very young readers (which she did well in more than fifty books). But the story strikes me as a great example of how our minds actually work. We rely mostly on immediate, sensory experiences for our actions. Our brains can handle abstractions but encode much better when there's a physical element attached. The intrepid little boy in the story can quickly encode "a flight of stairs" as he's racing down one, but it's harder to remember the need for pears until later in the story when he sees them on a farmstand.

The extensive and sometimes chaotic input we get from all our senses helps determine how happy we are moment to moment and day to day. I can be blissful in Paris, and my friend finds joy when she visits her noodle shop, but those special places are just that—special. Once we know the power of our senses, we can use them in practical ways to improve the joy we take in everyday life. What is outside of us changes our perceptions, but how about the inside? My next goal was to understand how our own bodies work to make us happy—or not.

PART THREE

The Unexpected Power of Sex, Exercise, and Diet

Certain physical activities, from taking a walk to having sex, are body enhancers—they make us feel more positive. Other things we do with (or to) our bodies like weight-loss diets don't make us happier at all—they are body deniers. Now's the time to start adding the good stuff that leads to great happiness and gratitude.

8

What Body Positivity Really Means

> There is, I came to realize, what the mind wants and what the body wants. The mind can excite the body, but its desires can also be false; whereas the body, the animal, wants what it wants.
>
> —CLAIRE MESSUD, *THE WOMAN UPSTAIRS*

IF YOU SPEND ANY TIME on Instagram or other social media, you've seen the hashtag #BoPo bandied about. Women and men of all sizes and shapes have gotten on the body positivity bandwagon, insisting that all bodies are beautiful and we need to reject the impossible standards of slim perfection set by society and media. The idea began as part of a fat activism movement in the 1960s then morphed into a wider realm. Variations on body positivity soon became a mantra among organizations trying to appeal to women (and occasionally men). You're beautiful as you are. You're enough. Love your body.

The hashtag may work well on social media, but being positive about your body only in terms of how it *looks* seems to be dramatically missing the point. Body positivity makes sense once it's seen in a much broader view than the everyone-is-beautiful mantra. Whether you celebrate bulging thighs, postpartum stretch marks, or rolls at your tummy isn't the point. The superficiality of beauty is only one tiny part of the information our bodies give to the world and what they take back.

The Dove company has spent a vast fortune on its Real Beauty campaign—a kind of do-goodism that hasn't done much good at all. Though meant to celebrate bodies of all shapes, it seems to suggest that when it comes to your body, shape is all that matters. In one of the early Dove ads, women had to walk through one of two doors marked "Beautiful" and "Average." Those who walked through "average" were told later that they had probably made the wrong choice. You should think of yourself as beautiful! Aside from the fact that most of us *are* average (that's what "average" means), should beautiful and average be the only choices women have for defining themselves? Maybe some women would like to choose the door that says kind or smart or funny or a good friend—but those aren't options.

Even when it comes to your physical self, the beautiful/average duality is offensively reductionist. You don't have to call yourself "beautiful" to appreciate that you have a body that's strong, healthy, fit, and competent and does what it's supposed to. In an effort to recognize that your body has functions beyond how it looks, some therapists began suggesting replacing body positivity with body neutrality. If you feel silly looking in the mirror and telling yourself that you're beautiful, then stop worrying about the mirror at all. The advocates of body neutrality suggest that you don't have to be positive or negative about your body—just accept it and don't think about it. Be dispassionate about your physical self because your body is not who you are.

But wait a minute! Of course your body is you—who else would it be? Developing a more accepting view of our physical selves shouldn't mean plunging backward to Descartes's view of mind and body as separate entities. It has taken science and philosophy a few hundred years to move from that mistaken approach and see mind and body as one. You can't ignore the physical part of you that is working with your brain every millisecond of the day to provide information and create the you who is you.

You want a body that interacts with the environment in a way to make you function better in the areas that matter for your life. We experience body positivity when we understand what our bodies are trying to do—and we can use the sensory messages of touch, sex, and current experiences to make ourselves happier. Let's try walking through *that* door instead, right now.

Current You and Future You

Most of the time, the human body gets it right. We need to reproduce to keep the species going, and our bodies are geared to want sex. We require calories to survive, and we find eating pleasurable. Exercise is good for our cardiovascular system and general health, and we feel good when we move vigorously.

At least that's how it should work. But your well-functioning body exists in a complicated world of social, cultural, and economic pressures that can interfere with an otherwise well-calibrated system. If you're doing tequila shots at a bar with friends, you're tuned in to the noise, the fun, and the social connection. Those can drown out your body's announcing that one more drop of alcohol will lead to a night kneeling in front of a toilet.

One question that keeps coming back to me is why our bodies do things that *don't* make us happy. We drink too much (sometimes) and eat too much (often) and forget to move or exercise. If our bodies are so smart, why aren't they keeping us on track and adding joy to our lives every day? The problem seems to be that your body lives in the present moment. When you touch, taste, hear, see, or smell, your body responds to stimuli occurring right now. It doesn't anticipate the future. Your brilliant body takes in the sensory experiences all around you at every moment, as many as it can, then sends all the details off to the brain to interpret. Your body doesn't bother with putting anything in context or even passing judgment.

We are often told that human beings are the only creatures that can actually plan for the future. I don't know if this is true—most dogs I know seem pretty smart about planning for a walk the moment their owners walk in the door—but it is the case that our ancestors figured out how to plant seeds, wait patiently, and have food many months later. This discovery of agriculture certainly changed humanity for the better. But that present-future understanding hasn't infiltrated many other areas of our lives.

We may be triumphant at planting now and getting food later, but with many other activities, we rely on our senses, our momentary experiences, to tell us how to behave and what to do. The body's one basic goal is to enjoy the moment right now. Planning for the future has many advantages, but your body wants to make sure that you survive into that future. Your brain follows the body's lead in this and doesn't bother to override the directive for immediate pleasure and survival. The future? Um, sure. Let's think about it tomorrow.

If you have trouble saving for retirement or resisting the piece of chocolate cake in the refrigerator, you can stop feeling guilty. Rather than suffering from a personal or moral failing, you are simply following the body's directive to make the most of the moment. "Gather ye rosebuds while ye may," as Robin Williams famously quoted in *Dead Poets Society*. Your body wants to grab the resources available in the present time. There's no sense thinking about tomorrow if you may never get there.

This live-for-the-moment spirit made sense in an earlier age of scarcity, and while you'd think our minds would have adapted to our newer and more plentiful circumstances, they haven't. Evolution moves slowly, and our bodies are still gathering the rosebuds while they can without much attention to what happens later. When we think about our future self—the person we'll be in two or five or ten years—we might as well be talking about a stranger. To your live-for-the-moment body, your current self and your future self can seem largely disconnected. If you ask someone you know whether they want $100 right now or $120 next year, just about everyone will prefer the immediate payout. In various experiments testing how we perceive the future, the same results show up over and over: *I'll take it now, thank you!* Economists find this outcome both frustrating and baffling. Unless you're getting 20 percent interest in your savings account (and good luck with that), the $120 payout next year makes much more financial sense. But all of us would prefer to get that $100 bill in our grubby hands right now. "Carpe diem," as Robin Williams also said in that movie. Seize the day.

At a party recently, I chatted with a group of people in their late thirties who had graduated from business school together about a decade earlier. They listened carefully as their friend Jared described a big career move he was planning. One of the classmates suggested that the plan had great immediate potential but major risks down the road.

"I've thought about the risks—but that's for Future Jared to deal with, and screw him," he said with a grin. (His language was slightly saltier.)

Everybody laughed, but Jared wasn't completely joking. Neuroimaging studies show that very different parts of the brain light up when people are asked to think about themselves right now versus in the future. The distinctions are so dramatic that the neural activity associated with the future self is very similar to what occurs when you're thinking about someone else entirely. For the guy I met at the party, the career payoff in his present-day registered as viscerally exciting. Future Jared didn't display in his body or brain as being any more compelling than the waiter across the room.

Harvard psychologist Dan Gilbert found that if you ask people how much they will change in the next decade, they don't expect much transformation at all. They assume they will like the same music and movies, have the same personality, and keep similar friends. But looking back on the previous decade, they usually report having changed a lot. This disconnect holds at every age. Whether you're twenty, forty, or sixty, you can see how far you've come but have no insight that the evolution will continue. How you are now is how you assume you will always be. Future You is unimaginable.

Your Body Wants to Seize the Day

Despite our indifference to our future selves, we're not necessarily making the most of our current selves either. Popular Australian philosopher Roman Krznaric says that the whole idea of carpe diem has been hijacked in what he calls the "existential crime of the century." When Roman poet Horace used the phrase two thousand years ago in his Odes, he probably intended a gentler suggestion than the "Just Do It" aggressiveness it's come to mean. Horace was mostly interested in plucking the ripeness from a fresh day and recognizing opportunities and joys as they come. Now we try to seize the day with a credit card (one-click shopping!) or what Krznaric describes as "a hyperindividualistic YOLO ("you only live once") mentality that places value on fleeting novelty and thrill-seeking above all else."

We also undermine our current selves by living at a distance from the sensual experiences of real life and substituting the remote events that get mediated through screens. I thought about that the other day when I finished watching the sixth season of a popular TV series a friend had raved about a few months earlier. (It happened to be *Gilmore Girls*, but don't get judgmental—you probably have your own hidden pleasure.) I streamed episodes while I worked out on the treadmill or stationary bike, and on nights when I happened to be home alone and feeling lazy, I watched while I lounged around. Before launching into the final season, I pulled out my calculator. The show had previously been on network TV, so there were 21 episodes each season of 44 minutes each. Some quick arithmetic showed that I had already devoted 5,544 minutes of my life to the series—the equivalent of almost *four full days*.

The number shocked me, but it really shouldn't have been a surprise. Most people in America watch about three hours of television a day, whether on a computer, phone, flat-screen, or other device. (Screen time spent on websites, social media, video games, email, and texts adds up to even more.) If you live into your eighties, you'll have used up about *ten years* on television alone.

Ten years is a lot of time. When your grandchildren ask you to tell them about the great experiences of your life, you're unlikely to mention your favorite episode of *Seinfeld* or even (snobbishly) *The Sopranos*, *The Wire*, or *Breaking Bad*. Our strongest memories are lived in person, with our whole bodies. You might smile or cry while watching TV, but to feel genuine joy, memorable emotion, or deep gratitude, we need a fuller experience than comes across on a screen. Or as Krznaric says, "If I were to ask you to seize the day, right here and now, it is unlikely that you would switch on the television and start flicking through the channels."

So what would you do instead to seize the day? Poet Andrew Marvell made a very compelling case that the best way to outrun time is with sensual pleasure. In his great seventeenth-century verse "To His Coy Mistress, To Make Much of Time," the poet explains to his beloved in glorious detail how he would like to spend hundreds of years admiring and praising her—he would devote two hundred years "to adore each breast" and "thirty thousand to the rest." But then comes the great crescendo explaining that he can't do it! Time on earth is too brief! Soon they will die and her long-preserved virginity "will turn to dust /And into ashes all my lust." They should indulge in each other before it is too late.

I'm told that boys at English boarding school used to

memorize Marvell to recite on dates, and it is surely one of the finer seduction tools ever written. I remember reading the poem for the first time in high school and thinking it made absolute sense. (For what it's worth, it took many years for me to act on the conviction.) Marvell's passion thoroughly convinced me that we needed to fill our lives with amorous fervor. Carpe diem! Indulge in earthly physical pleasures right now! Marvell's may be the greatest carpe diem argument of all time.

The Pleasure of Touch

For most of us, our bodies crave the pleasures of touch, sensuality, and sex, and disconnected from any other demands, they would happily follow Marvell's seductive advice. But we don't always give our bodies the chance to make us happy. Krznaric worries that what he calls "the legacy of Greco-Roman moral ideals and hair-shirt Christian teachings" have made physical joy seem wrong. "For two thousand years there has been a long war against pleasure," he says.

How do we end the war and regain the experiences our bodies want in order to make us happy? If we really want to be positive about our bodies, we need to recognize the physicality they crave and the simple sensory experiences that can make us happy or provide pleasure. Even our language reveals our yearning for sensory connection. When you hope to see someone again, you say "Let's be in touch," and when you've gone too long without calling a friend, you apologize for being out of touch. You were also out of seeing, smelling, and hearing—but those don't trip off the tongue in the same way.

Back in the 1960s, a psychologist at the University of Florida decided to study how often couples touch each other when they are talking together in coffee shops. He made his observations in several parts of the world and reported results from a high of 180 touches per hour in San Juan, Puerto Rico, to a low of 0 touches in London. His hometown of Gainesville, Florida, came in with a paltry two. France was somewhere in between with 110. The kinds of touches also varied, with most being short contact to the hands and shoulders—except in Puerto Rico, where he observed longer connections that extended to the face and arms. That our physical interactions come with cross-cultural distinctions makes intrinsic sense, even if the numbers seem a little off. Touching 180 times an hour equals 3 touches every minute, which sounds more like an adult film than a coffee shop conversation.

Nonetheless, I'd be happy to go to San Juan, because when I'm sitting with someone I like, my instinct is to touch their arm, rub a hand, or squeeze a shoulder. I mostly resist these days because touching has become increasingly verboten in America. An older friend told me how uneasy he felt at work one day when a distraught colleague tearfully told him that her dad had just died. His natural instinct was to reach out and offer a hug of sympathy or a shoulder to cry on. But in a litigious environment where even complimenting someone's outfit can bring charges of harassment, he didn't dare.

Our touch-phobic society might have evolved with good intentions, but it also interferes with the very positive effects that come with sensory connections. I'm all in favor of respecting your body and everyone else's, but most bodies yearn for touch. Respecting your body doesn't mean encasing it in glass. The best physical experiences change our emotions and inspire feelings of gratitude, joy, and connectedness, and one of the simplest and most powerful expressions of comfort and connection comes from touching. Patting a child on the back (now banned in many school districts) can encourage them to be more confident in a presentation or feel more positive about reading. Studies going back decades show that babies crave physical closeness and those raised in institutions and deprived of touch suffer serious developmental challenges, even if all their other physical needs are met. (Hug your children early and often.)

One of the saddest studies of the power of touch occurred inadvertently under the dictator Nicolae Ceaușescu in Romania in the 1980s. Under his horrific orders, which curtailed abortion and contraception and sent a generation of abandoned children to be raised by the state, thousands of babies ended up in orphanages without enough people to care for them. While some were treated cruelly, others were fed sufficiently but otherwise left alone. Nobody touched them or hugged them. Parents in America and around the world who later adopted these babies found they couldn't overcome this early touch deprivation. The babies developed emotional problems including bipolar disorder and schizophrenia—and even more surprising, they suffered from physical problems including weakened immune systems.

We've known about the power of touch for a long time—and it's a huge mistake to ignore it. Back in the seventeenth century, the philosopher Johann Gottfried von Herder pointed out that a young child in a nursery is constantly gathering experiences by "grasping, lifting, weighing, touching, and measuring things with both hand and foot," learning more that way "than ten thousand years of mere gaping and verbal explanations could provide."

When someone touches you gently, you're more likely to cooperate with them or agree to a request they make. Touching lowers stress levels and causes the release of hormones connected to happy feelings. Touching someone who is sad or in pain has soothing effects that words can't match. One study found that, even separated by a barrier so they couldn't see each other, people were able to communicate emotions like compassion, gratitude, and fear with a very brief touch to the other person's forearm. Our bodies are so sensitized that you can quite literally feel someone's compassion when they touch you.

Touch has an intrinsic value unrelated to any social connections. It can be hard to untangle the physical effects of the touching from the social and emotional connection that sometimes comes with it, but the two experiences can exist separately. In one study, a group of older people got regular visits from someone who provided conversation and good cheer and another group got the same visits with massage added on. The second group reported greater emotional benefits, and they showed more cognitive improvements too. The touch creates a positive experience on its own.

The Imprint of Your Fingerprint

Your body comes prepared to experience the world through sensory touch, with thousands of sensors in each of your fingertips so sensitive that they can pick up details of an object half the size of a human hair. These thousands of sensory receptors each activate a neuron in what's called a receptive field. In the fingertips, these fields are densely packed—each field covering barely one-tenth of an inch. To get an idea what this means, take a couple of sharp pencils and hold them fairly close to each other on the back of your wrist or on a fingertip. (Or ask a friend to do it—this is a good game to play with someone you're trying to get to know.) How far apart do they have to be before you can sense two separate points? Each neuron in the fingertips covers such a small area that you can feel a distinction when the points are fairly close. Now try the same thing on a part of your body like your trunk or legs where there are a lot fewer receptors. You (or your friend) will have to hold the pencils much farther apart, maybe an inch or more, to know there are two separate points.

Many toy companies produce a version of a game where

toddlers reach their little hands into a box filled with objects and touch and squeeze and feel what's inside. The information on the packaging usually boasts of great educational value. But the truth is that while the game may be fun, we don't have to teach a toddler's sensory neurons how to fire. The process comes naturally. The neurons in the hand and fingers send the brain an accurate representation of what is being touched or felt without any conscious interference necessary. As usual, it's the brain that needs some time to catch up and process. Baby's fingertips send the image for "fuzzy ball" long before the brain knows what to do with the information.

Researchers have long figured that the whorled ridges of our fingerprints must serve some purpose other than identifying killers on *CSI*—but the exact purpose hasn't been clear. New evidence suggests these whorls are linked to particular sensory fields. The distance between the nerve fibers that connect to the receptive fields in the fingertips closely matches the lines and ridges of the fingerprint. This amazing ability of your body allows you to recognize minuscule differences in texture or pressure or shape.

Our brains seem to have two pathways for processing the sensory information that your fingertips, hands, and other sensory receptors are constantly sending. In a similar pattern to how pain gets processed, the sensory information first goes to a part of the brain that processes it for straightforward particulars, letting you know (without looking) whether you're touching a prickly cactus or a velvety flower. The sensory neurons report distinctions of texture, temperature, location, pressure, and shape without adding any emotional overlay. A second sensory pathway goes to the parts of the brain that process social and emotional information—which is why the gentle touch on your arm from a lover feels very different from the same touch from your pushy boss.

Our sense of touch is fundamental to who we are and has been described as offering us "our securest and most reliable knowledge of the external world." Touch has a slowness and intimacy that make it more potent than the colder and quicker perceptions we gain through sight. Our awareness of where we are in the world and perhaps of *who* we are comes largely through the sensations of our bodies. That's another reason why the "body neutrality" movement seems to miss the point. You don't want to be neutral about your physical experiences, because they define so much of your world.

The Softest Skin

I'm not much of an impulse shopper, but if I'm in a store that has a fancy display of hand creams, I can't resist. My nightstand fills up with various lotions that I use morning and night because I like my hands to feel soft, and I have small tubes of hand creams stuffed into my pocketbook to use during the day.

Most of us spend more time fussing over our skin than any other part of our body, which makes sense, since skin is the largest part of our body. Americans spend some $20 billion annually on skincare products that will make us glow, gleam, and shimmer. I understand that a lot of that outlay comes with a focus on beauty, but we also care deeply about how our skin feels. Given all those sensory receptors, we are aware when our skin feels rough or tingly or excessively dry. That soft skin may even allow our sensory neurons to be more responsive—though it's probably not necessary to spend the $1,090 for one ounce of La Prairie to get the best effects.

The pleasures of touch can come from the simplest sources. The other day, I was interviewing a well-known celebrity when her dog ran into the room and jumped up on the sofa next to me. The celeb apologetically tried to grab the dog away, but the lovely Rummie (a King Charles spaniel) was busy licking my arms and hands and wouldn't be distracted. I gamely began rubbing his back, and soon he snuggled against my leg and fell asleep. For the rest of the nearly three-hour interview, I kept one hand on his soft furry back, gently petting him over and over. At one point, Rummie woke up and looked expectantly at his owner, and I realized I didn't want him to leave my side for his more familiar friend. The sensory experience of touching his soft, warm fur was keeping me happy and relaxed.

Rummie (like most dogs) knew he wanted to be petted. I didn't realize that my body yearned for the same gentle touch—but it did. Would rubbing a soft furry rug or a piece of velvet have had the same calming effect? Probably to some extent—but not fully. Our senses can pick up emotional vibes well before they get to our brain. The nerves in my fingertips picked up the warmth of the dog's body and the slight changes in his movements. The message to the neurons had more information than my conscious brain realized, and it created an automatic relaxation response. My blood pressure lowered as did all my cortisol levels. Had Rummie tensed or suddenly prepared to attack, my body would have known it well before any response registered in my mind.

Some hospitals now have programs where volunteer therapy dogs (and their owners) come to in-patient floors where they're brought room-to-room to interact with patients who express interest. Patting and hugging the furry animals regularly reduces stress levels and anxiety. Dale Needham, a professor of medicine at Johns Hopkins, has even shown the benefits of bringing the dogs into intensive care units where, he says, doctors should think about relying less on medicine and more on nonpharmaceutical interventions including "animal-assisted therapy."

In other words, the power of touch is very powerful.

The Joys of Sex

If simple touching changes our bodies and emotions, how about more dramatic physical engagement? In reading many reports and research studies on happiness and well-being, I noticed how rarely sex was mentioned. Given how much time people spend thinking about sex, engaging in sex, and thinking about how to engage in more sex, the activity is clearly important to our happiness and sense of pleasure in the world.

Your body craves sexual contact and does everything possible to make it a happy experience. Social, cultural, and religious obstacles may get in the way of your mind embracing a sexual encounter, and in this case, finding a truce between body and mind isn't always easy. Generations of young men and women who were taught that masturbation is harmful but found themselves hiding under the bedsheets indulging anyway can attest that the body wants what the body wants.

When you have an orgasm, dopamine, oxytocin, norepinephrine, and serotonin all flood through your body—a potent cocktail that makes you happy and calm. Your mood soars upward while your stress response goes down, and the relaxation makes it easier to fall asleep. Your body sends the message that sex is desirable and you should seek it more and more.

If you ever watched the Showtime TV series *Masters of Sex*, you know that a lot of things interfere with your body's desire for physical gratification. William Masters and Virginia Johnson began their studies of human sexuality in the 1960s and produced some of the first-ever clinical data on the human sexual response. They found, among other things, that there's no difference between a vaginal or clitoral orgasm (the nerve pathways are all connected) and that both men and women can continue to have strong sexual responses as they age. (Later studies have strongly supported both findings.) Their books became bestsellers, and the information should have been the start of more research interest, but the medical community found Masters and Johnson's work scandalous.

"Sexual satiation is as instinctive as the urge to sleep," Dr. Masters explains calmly in one episode.

"Sleep doesn't have the potential to be the ruination of our moral code," says an outraged colleague.

I got in touch with Nicole Cirino who heads the reproductive psychiatry department at Baylor College of Medicine in Texas. With degrees in both ob-gyn and psychiatry, she has a fresh perspective on how sex, hormones, and reproductive cycles affect our happiness and well-being. She told me there's not much question that "sex is good for us and increases our well-being." Exactly how that works is still being studied. Some researchers attribute the mood-boosting effects of sex to the intimacy and closeness that often come with it, but there's good evidence that the effect can flow the other way too. The physical closeness leads to an emotional connection.

On the simplest level, the sexual response model has a few steps that Masters and Johnson first outlined—excitement followed by arousal, an orgasm, and finally a period of sexual satiety. If you like the person you're with (and we're talking only about consensual sex here), the physical pleasures allow for emotional bonding. But long before your conscious brain gets in on the act, your body is sending its own signals. Cirino says that when you have an orgasm, you experience an increase in "serotonin activity, endocannabinoid activity, and endogenous opioids." All of those are chemicals and neurotransmitters that increase our happiness and pleasure.

A little more information about those chemicals makes the power of sex even clearer. Endocannabinoids were first identified in the brain in the 1990s, and they are structurally similar to the active ingredient in cannabis. ("Endo" or "endogenous" just means it's made by the body.) The first one discovered was named *anandamide* from the Sanskrit word for *"bliss."* The opioids reduce pain and modulate emotions. Essentially, after sex, the body naturally produces chemicals that make you calm and happy and less aware of pain. Other pleasure-inducing chemicals including oxytocin, dopamine, and endorphins also surge during sex, putting you in a calmer, less stressed, more euphoric mood.

From an evolutionary standpoint, it makes absolute sense that sex comes with a flood of body-made chemicals that leave you slightly elated. The positive feelings make you want more sex—and the more sex you crave, the more likely it is that the species will survive. One study out of Essen, Germany, even found that the immune system gets activated during sexual arousal and orgasm. The number of white blood cells (which fight infection) increases, so it's possible that having sex makes you healthier. (An orgasm a day keeps the doctor away, as one headline put it.)

When you're deep in the throes of sexual desire—and certainly in the midst of an orgasm—the hormonal activity is powerful and controlling. Your brain may try to throw out a thought now and then, but your body is the main player. Perhaps that helps explain why the major religions have placed such extreme restrictions around all kinds of sex, forbidding even the delights of self-pleasure. The men who inscribed the rules a couple of thousand years ago didn't know about oxytocin and endorphins, but they did know about wanting to maintain external control of people and not allow the body to lead the way. They made sure sex was about propagation, not pleasure.

Having sex with a partner adds sensory touch to the physical pleasures and may make the experience even more intense, but your body isn't all that particular. Those calming and happy-making chemicals and neurotransmitters get released after orgasm no matter how you get to the climax. You can create the neurochemical changes for happiness and improved mood on your own.

Why Captain America Gave Up Sex

Our bodies want us to engage in as much sex as possible. If our bodies had their way, we'd have sex regularly, with or without a partner, to enjoy the physical thrills that accompany it. But our religious, social, and cultural reins are always pulling back on the pleasure. Sex may be the most natural activity that a body can experience—none of us would exist without it—but cultures go through different cycles, and America has slipped back into an era where sex can inspire more retribution than joy. I would call it puritanical except that the Puritans liked sex a lot. They celebrated sex within marriage, and while the laws didn't support premarital or extramarital sex, historians studying court and church records say that both were rampant in New England in the seventeenth and eighteenth centuries. Americans now, in contrast, are having less sex than ever, and nobody is exactly sure why. Some researchers have suggested that people get distracted by technology and social media or that the fallout of the #MeToo movement has made all sexual encounters seem tainted.

The layers of uncertainty about what is acceptable means that while romance thrives in books and big-budget studio movies, sex often gets treated timidly or ignored altogether. The characters may look sexy but are devoid of lust. A movie can stay PG-13 safe with bodies being slashed or beaten but not when those same bodies provoke gasps of pleasure rather than pain. In an article for *Time* magazine, writer Eliana Dockterman pointed out that the comic-book superheroes who now dominate in movie theaters are allowed to be violent but not sexual. Nobody ever takes off their form-fitting costumes, and it's gotten to the point where Captain America, played by the very handsome actor Chris Evans, "never ventured beyond a brief kiss in more than half a dozen Marvel flicks."

Removing sex from our lives and screens changes the basics of being human. Academy Award–winning writer and director Edward Zwick, known for his stirring movies and TV shows, joked that taking away lust makes the heroes even less human. "To not include sexuality is to take one whole color off your palette," he said. "You end up with relationships that are less complex."

Even studying the link between sex and well-being makes some people nervous, so the field hasn't been growing very much. Nicole Cirino came to Baylor to build their sexual medicine program, but that line of investigation has been sidelined in favor of perinatal mental health. She now spends most of her time on the issues surrounding pregnancy and childbirth.

I asked Cirino if sexual desire is a key concern for the couples who consult her, and she laughed. "Often it's the last thing they'll ask about!" she said. Many are too overwhelmed with the challenges of daycare and the anxieties of work stress and career-life balance even to think about sexual function. And after months of pregnancy, childbirth, breastfeeding, and the exhaustion and tension that come with having a newborn, many couples have stopped having sex for months. "The couple has to really consciously reengage because it's not going to spontaneously happen," she says. But getting the positive momentum back allows your body to start improving your situation too.

In the Pulitzer Prize-winning novel *Foreign Affairs*, written many years ago by the late Alison Lurie, a small and seemingly prim female academic in her fifties begins having a passionate affair with a large and loud cowboy named Chuck, who doesn't seem at all her type. But she likes sex and, in her quiet way, has always found a way to make it part of her repertoire. She's not sure if Chuck loves her (it turns out he does—very much), but she doesn't think it matters. "Physical pleasure of the sort she's known with Chuck does improve the entire world," she thinks. "It becomes a humming, spinning top in which all the discordant colors are blurred and whirled into a harmony that spirals out from that center."

Your mind doesn't always know what's going to improve your well-being, but give your body a chance, and it does it right. Adults who have sex one day report feeling happier and finding greater meaning in their life the next day. (It doesn't work in reverse—feeling happy one day doesn't mean you'll have sex the next.) The postcoital glow that gets joked about is very real. Sex increases blood flow and sends feel-good chemicals circulating in your system. It also helps you relax and sleep better, and it improves your overall health, happiness, and general quality of life. Should we have more sex? I'll leave that to you. This is one case where your body wants what your body wants—but it can't always get it.

Sex may be a source of joy that you need to use with caution—but using it will have consistently positive results. As Masters and Johnson pointed out, you don't have to lose the pleasures of sex at any age or in virtually any circumstances. Popular images in advertising and media suggest that only the young and beautiful have sex, but it remains a universal pleasure. One sunny day when my husband and I were in our midforties, we were walking through a pretty park and he leaned over and kissed me. Warmed by both the sunshine and the pleasure of closeness, I put my arms around him to continue the affection. A couple of teenagers walking by groaned in disgust when they saw us. "Eew, old people kissing," one of them said loudly. We started laughing so hard that all romance disappeared. In the many years that have gone by since then, "Eew, old people…" has become a private joke that we repeat often. I'm happy to report that instead of being a discouragement, it regularly inspires us to continue kissing.

Listening to what our bodies want—however they appear to others—can bring a new sense of comfort and joy. Sex is one of the most basic physical instincts, and eating is another. How is it that we have made these two fundamental pleasures battlegrounds for conflict and embarrassment? I started wondering why our bodies and minds can't always agree about what we should eat—and how we can bring some sanity back to our diet-obsessed culture. I was surprised by what I would discover about food, bodies, and happiness.

9

The Happy Body Food Plan

> Moderation. Small helpings. Sample a little bit of everything. These are the secrets of happiness.
>
> —JULIA CHILD

A *NEW YORKER* CARTOON THAT always makes me laugh shows a woman on an exercise mat diligently doing squats with weights. On the mat next to her, another woman is sprawled out, eating a bag of chips, with the caption "What? The teacher *just* said, 'Listen to your body.'"

Yes, you feel happier and get healthier when you exercise, but the body-mind partnership doesn't always work as smoothly as we'd like. Most of us can fully sympathize with the woman who thinks her body wants her to loll around with a nice bag of chips. (I'd prefer chocolate.)

Certain foods do make us happy. I've asked many people what they'd put on the menu for their last meal and gotten answers from mushroom pizza to filet mignon to blood sausage (a friend from Munich). One person chose to forgo the main course altogether and request an ice cream sundae. While people's answers differed dramatically, everyone I spoke to got a little smile on their face at the question, followed by that faraway look as they envisioned the food that would make them most happy and content.

Our bodies send our brains positive reinforcement for the foods that will keep us healthy and happy. But those reward centers were established long, long ago and haven't quite caught up. Salt used to be hard to come by, which is why all animals look for it. When your dog licks your sweaty arm, he may love you, but he mostly wants the salt. The food writer Mark Bittman points out that once humans began to mine salt, it was very plentiful, but "You don't stop having Stone Age impulses just because they no longer serve you in a modern world." When our cartoon lady thinks her body is telling her to eat chips, she may be responding to a deeply wired pleasure center for salty foods, now manipulated by food manufacturers to make you crave something we no longer really need.

As with so many of the signals we've already encountered—from feeling back pain to tasting wine—our body's messages get filtered through a brain that can interpret them in many different ways. It's a little like listening to Fox News, CNN, and MSNBC on election night. They're all getting the same data, but they're going to tell you very different things about what they mean.

All the social and cultural noise that influences our

conscious minds makes it harder to hear the body's messages. Consider that some thirty-five thousand people gather every July 4th at Coney Island to watch the competitors in Nathan's Famous International Hot Dog Eating Contest. The current world record is seventy-six hot dogs and buns swallowed in ten minutes. An organization called Major League Eating (I'm not making this up) keeps official records of how much champions have devoured in other competitions around the country, involving an extremely long list of foods including pizza, chili, apple pie, shrimp, grits, grapes, fruitcake, crab cakes, baked beans, tacos, doughnuts, pickles, pasta, watermelon, turkey, burritos, and asparagus (the winner ate twelve pounds in ten minutes—and they were deep-fried). I hesitate to mention additional records that involve mayonnaise and butter, since the very thought makes me queasy.

That's the point, of course. Your body does not want you eating sixteen pints of vanilla ice cream in six minutes (the record at the Indiana State Fair in 2017) or thirty-four cannoli. These are extremes, but just about all of us have stood up from a holiday dinner and moaned "I ate too much" or "I'm so stuffed" at some time or another. Had you put down your fork halfway through the main course and waited a few minutes, your body would probably have let you know that it wasn't hungry anymore. On the other hand, heeding that announcement would mean ignoring the copious dessert table laden with pumpkin, apple, and cherry pies and a homemade chocolate mousse that everyone around you declared delicious.

My older son is one of those rare people who actually do put down their forks and listen to their body. He takes small portions at dinnertime and eats slowly. When I ask if he wants more, he invariably says he'll decide in a few minutes. He's already aware that Mom's roast chicken is delicious (trust me, it is), but now he's listening for a physical signal that tells him whether he needs another helping. Most of us don't do that. It takes a little time for the digestive hormones in the stomach to send a signal to the brain about satiety, but most of us are well into our second helping before those hormones start talking.

Most popular weight-loss diets take the exact opposite approach. They provide a rigid eating plan that has absolutely nothing to do with what your body is actually demanding. If we were listening to our bodies, would the cabbage diet have ever become popular? The seven-day detox with lemon juice and cayenne pepper? There is no population group on earth that eats only cabbage or lemon juice as a normal, natural diet—so why would we possibly consider either of those a reasonable way to eat? As Mark Bittman puts it: "Balance is good. Imbalance is bad. Really. Period."

Awareness of our bodies goes a long way. When I spoke to Judson Brewer, the Brown University expert in behavior and mindfulness (who we met in Part One), he told me, "The only way to change a behavior is to pay attention. I love how simple that is. It's nothing about the myth of willpower—it's about seeing the reward value go below zero."

A lot of research shows that our highly processed foods

are aimed at hitting the pleasure centers of the brain, which makes it harder to resist consuming too much. Brewer thinks that while sugar and fat are pleasurable, "your body is able to recognize when it's too much. If you're paying attention, you'll stop eating."

Brewer has created various eating apps to help people ask questions of themselves and know what their body is actually saying when they experience various food cravings. Could it be that you're just feeling restless? How recently have you eaten? Is your stomach growling? If you just ate thirty minutes ago, there's a good chance that your body is feeling something other than hunger. The apps help you differentiate between stress, boredom, and hunger so you can make the proper response.

"Isn't it ironic that you're using technology to help people listen to their own bodies?" I asked.

"Yes, ironic," he said. "At the beginning, having a neutral arbiter can be helpful, but then we want the apps to be deleted. At some point, you hear it when your wise body says 'Dude, what are you doing? Overeating doesn't feel good.'"

The Pancakes and French Toast Conundrum

When I invited some friends over for dinner the other night, I made squash soup, Caesar salad, roasted salmon, rice pilaf, and asparagus, and then I brought out a particularly luscious cheesecake for dessert. One of the men at the table groaned that he'd already eaten too much—but then cut himself a large slice.

"Eating cheesecake has nothing to do with being full," he said.

We all laughed, but his comment had some ring of truth. My dad used to joke that you have one stomach for dinner and one for dessert—and I believed him for much longer than I care to admit. In sixth grade, we learned that cows have four stomachs for digesting their food, and I lost points on a test by saying they had twice as many stomachs as humans. I'm not sure the teacher fully understood that it was more than my math that was wrong.

While my understanding of human anatomy might have been mistaken, I was right in thinking that our bodies can differentiate between dinner and dessert. Humans have one stomach and a single digestive system, but it's completely true that our bodies can be sated with one kind of food but eager and prepared to eat another. Researchers call it sensory-specific satiety. Your body is always looking for new and different tastes and textures—a wise move, since a range of foods will provide more nutritional variety. You can test this yourself. Eat one or two apples and you're probably not eager for another one, but if a pear or pineapple is available you might be happy to keep eating. That's why buffets can be so wonderfully problematic. You see all the possibilities—French toast, pancakes, eggs, sausage, bacon, and you want all of it.

Chefs in the fanciest restaurants like Per Se in New York City or the French Laundry in Yountville, California, take advantage of our pleasure in new flavors with multi-course tasting menus that cost hundreds of dollars. You get just a bite or two of an exquisitely prepared dish, and before your senses can get accustomed, it's whisked away and replaced by another. You experience a constant onslaught of sensory pleasures—a perfect example of your body making you happy. It's reasonable to wonder if the top chefs get extra credit for their brilliance because nobody ever gets to eat so much of a dish that it loses its excitement.

Unless you have endless time and money, you're unlikely to go to Per Se or the French Laundry, but a couple of their approaches can work at home. Pierre Chandon, a chaired professor at the INSEAD-Sorbonne Behavioral Lab, told me that moving the focus from *what* we eat to *how* we eat (and how much) can make everyone happier. When we spoke, he had a bookshelf behind him filled with the memorabilia of his research—like the oversize soda cup from 7-Eleven, which provides good value but not much happiness. He also had a bust of Epicurus, the Greek philosopher who understood two thousand years ago that the wise person looks for pleasure in food, not quantity.

"Sensory pleasure peaks during the first few tastes and then diminishes as you continue," Chandon told me. "Consider eating a chocolate mousse. The first spoonful is the most delicious, where you get the WOW effect. The second bite is still tasty but less than the first. And the last one, if it's a large portion, makes you a bit queasy."

To get to the total pleasure you've taken in a food, your

mind doesn't add up the happiness in each bite but rather takes the average. Chandon says that eating an excess amount can actually *decrease* your pleasure because it draws the average down. Think of it this way: The first bite of that chocolate mousse may register as a 10 on a scale of 10. You're thrilled! What a treat! The next couple of bites may also be delicious, but then the sensory pleasure starts to diminish. Your body isn't paying as much attention because the taste is no longer new, so your brain also loses interest. By the time you're finishing what's on the plate, you may be feeling more dutiful than thrilled. The pleasure scale for each bite may look something like this:

10, 10, 10, 9, 9, 7, 6, 5, 4, 4, 3, 3

Average those together, and you are well below the joyous 10 you experienced in the early tastes. Our total reaction to an experience is heavily influenced by the ending—so when you finish a huge portion, you may even forget the greatest pleasures of the beginning. "The last spoonful doesn't add pleasure, and it draws the average down," Chandon said. "If you put pleasure at the center of what to eat, you'll share the dessert and not eat it all."

Chandon's advice makes perfect sense, but it goes against the grain of American eating habits. While the fanciest restaurants have been preparing smaller portions of each dish, most of the rest of America has been busy supersizing. The original Coca-Cola bottle was 6.5 ounces, and for decades, that was the only size you could get. Now that size would barely qualify as a toddler cup. Chandon points out that, even if you don't purchase the humongous soda cups sold in convenience stores and movie theaters, their availability changes our perception of what normal looks like. Similarly, if you go to a popular restaurant like the Cheesecake Factory, some of the main courses are so enormous that they clock in at over 2,000 calories. The Louisiana Chicken Pasta looks big enough to feed the whole state rather than a single person, and the cheesecake slices aren't a delicious taste—they're a 1,500-calorie blob. The point isn't that we are eating too many calories but that we have stopped thinking about the pleasure we take in each food. We've all been taught not to waste food—but the real waste is eating everything that's in front of us if it makes us less happy.

The Pleasure Principle

Shortly after I spoke with Chandon, I went out to dinner with friends, and for dessert, I ordered the tiramisu. I took a bite and it was delicious. Then another bite—equally good. I put down my fork and thought about it. The taste had given me the after-dinner burst of pleasure that Chandon described—but would eating more of it increase my pleasure? Probably not. I had to convince myself that it would be okay to leave the rest of the enormous slab on the plate. Yes, I'd paid for the whole thing, but eating it all wouldn't increase the value. The smaller portion offered the highest level of happiness I could expect.

Chandon told me that trying to educate people or

guilt-trip them into eating healthily never works. But trying to refocus on food making you happy seems to do the trick. Food should be a source of joy rather than a hub of guilt. Instead of talking about food in terms of calories and saturated fat and cholesterol and added sugar, reframe food as one of life's great pleasures. Think about what foods give you joy and how much of them you really need to get the ultimate enjoyment. Satisfying hunger is necessary to survive, but satisfying our sensual desires comes in a very different category.

I once produced a television series for the Food Network where celebrities took us into their home kitchens to cook their favorite meals. For a holiday-season show, former *Saturday Night Live* cast member Julia Sweeney prepared an amazing dinner of turkey, stuffing, mac and cheese, green beans, and other Thanksgiving classics. When we discussed what she wanted to make for dessert, she didn't hesitate—a triple-layer chocolate cake with chocolate frosting. That sounded like a heavy ending to a rich meal, so I suggested we go with something lighter—maybe a berry pie or fruit salad.

She shook her head and gave a firm no.

"If you don't have something chocolate, how do you know a meal is over?" she asked sweetly.

Julia always made me laugh, and that comment particularly stuck with me. Our bodies and minds take in all sorts of signals to decide how much we should eat. A taste of chocolate may be a good reminder that you're no longer hungry. It may also be the bit of pleasure your body needs.

In my own adaptation of Julia's example, I keep a chocolate bar in my desk drawer at home, and many evenings (when I haven't had chocolate cake for dessert) I take it out, break off a small piece, and sit down to savor its rich taste. Simply knowing that I am going to get this late-night pleasure gives me a tingle of satisfaction. I'll eat less at dinner knowing that I still have a sensory pleasure ahead. The chocolate ritual helps to calm the caveman instinct we all have—which is to eat everything possible when it's available because you never know what's ahead. Even in our age of plenty, it's hard to shake that off, but having my secret stash of chocolate lets me ignore the caveman in my head. I can stop eating dinner without panicking because I know there will be something ahead to please my senses. No guilt, just joy.

Sometimes, you don't even need to eat the food to get some pleasure out of it.

A Broadway actress I interviewed recently suggested we meet for coffee at a bakery called Lady M. I'd never been there before, and the moment I walked in, I was overwhelmed by the sumptuous cakes in the display case. As we sat down and looked at the luscious menu, I couldn't help launching into my own performance.

"*Lacy handmade crepes layered with delicate rose-kissed pastry cream. Finished with a glossy sweet rose jelly and edible rose petals,*" I said, reading one description aloud.

The actress laughed and immediately followed up with the next menu item.

"*Delicate raspberry, blackberry, and black currant mousse piped over a billowing mountain of whipped cream topped with baby pink macaron shells,*" she intoned in her throaty voice. Then she added, "Just reading about it is so satisfying, I'm not sure I even need to order!"

We did order, of course. We each got a slice of a different cake and took a forkful of the other's choice for double the pleasure. But the actress was right that the sensuous descriptions were enormously satisfying in themselves, a kind of mind-body theater. Chandon found that menu descriptions that give a highly descriptive and sensual focus make people eat more slowly and enjoy the meal more. When asked how much they would be willing to pay for the richly described meal, people offered a higher amount than those given the same meal with only cursory descriptions. Our mind and body work together to tell us how much pleasure we can take in what we eat.

Our bodies yearn for pleasure, and in the reverse of what you might think, admitting that we want something delicious can be a positive step. Americans have come to associate tastiness with unhealthy food, but it doesn't have to be that way. In studies in France, people have the opposite intuition, spontaneously associating healthy food with being satisfied and content. Describing healthy meals in terms of their taste and pleasure makes people enjoy them more—and eat them more regularly—than if only the health benefits are emphasized. Any parent can tell you how this works. Tell a child to eat carrots because they're healthy and you'll get the disgusted face that only a five-year-old can produce. Offer the same child a yummy, tasty carrot stick and there's some chance that they'll actually take it and like it.

Focusing on eating for pleasure can actually help you make healthier choices. In one study, people were given a granola bar and told either that it was a healthy snack with lots of good-for-you stuff or that it was a delicious chocolate and raspberry treat. Interestingly, people found the healthy bar *less* satisfying than the same bar that was described for pleasure.

"We tend to have a binary approach to food—we think it's either good or bad," Chandon said. Americans associate healthy food with being tasteless—jokes about kale are a gimme for stand-up comics—and so take less pleasure from the same bar when it's presented as being nutritious rather than delicious. Chandon has found that putting a "health halo" around certain foods backfires in yet another way since you'll simply end up overeating. In one study, he offered people a bowl of M&Ms labeled low-fat and low-calorie. While there is no such product (he explained the deception afterward), people thinking they were making the healthier choice ate 46 percent more than those offered the regular candies.

It reminded me of a time when low-fat diets were all the rage and magazine articles insisted that if you stayed away from fats, all would be fine. I had written about nutrition for many years by that point, but even I was convinced. I remember telling my friend Vicky that I ate Twizzlers every afternoon at work—a good choice because they were low-fat so I wouldn't gain weight. She looked at me and shook her head.

"You're lucky," she said with just a twinge of irony. "My body has actually figured out how to store sugar as fat."

The comment hit me like a sledgehammer. Well of course. We can label things however we want—but the body works from its own intelligence.

Some mind-boggling studies by Alia Crum, a psychologist at Stanford, show that what we think we're eating affects more than our emotional response—it changes our physical state too. In one groundbreaking study, she gave participants two different milkshakes. The first was a vanilla milkshake with a label showing that it had zero fat and sugar and a scant 140 calories. A week later, they returned for an indulgent shake with a label revealing 620 calories with lots of fat and sugar. Crum measured the amount of the hormone ghrelin circulating in the volunteers' systems after each shake. Ghrelin, known as the hunger hormone, gets secreted by the gut when your body needs more calories. It signals the brain that it's time to eat something, and it also slows the metabolism just in case you don't get food immediately. On the other side, when you eat a big meal, the ghrelin level drops—time to stop eating—and the metabolism revs up to digest what you've just devoured.

The ghrelin levels she recorded tracked very closely with the calories in each milkshake. They dropped considerably when people got the higher-calorie shake and stayed relatively flat for those who got the less filling 140-calorie shake. That sounds reasonable until you discover the twist. *Both of the shakes were exactly the same.* Each contained 380 calories and the only thing that differed about them was the labels.

This is pretty stunning. The volunteers thought they were getting different amounts of calories on the different weeks, but they weren't. Still, their bodies responded to what they *believed* they were consuming rather than what they actually did. Their reports of how full they felt coincided with what they thought they ate, which isn't all that surprising. But that the mind-body interplay is strong enough to change the release of a hunger hormone is far beyond what most of us would expect.

Crum had done a lot of previous research on placebos, so she might not have been quite so surprised. She understood that our mindset can change not just our emotions but our physiology. As she put it, "What we believe, what we expect, what we think about the foods we eat determines our body's response."

All this makes me wonder about those long-ago Twizzlers snacks. It's possible that if my mindset held that the Twizzlers were healthy, my body responded in kind—perhaps releasing or not releasing some hormone that matched the expectation. Nobody has done any experiments to suggest this—but it does fit in with what we are starting to understand about the power of the body to influence the mind and the mind to influence the body.

Should a researcher like to study the Twizzler Effect, I am available as a subject.

The Happy Body Lifestyle

One of the greatest happiness bugaboos for many people is the outward form their bodies take—the muscles and fat, the biceps and breasts, the shape and height. If you watch enough TikTok videos about nutrition and exercise, you start to think that all of it is under your control. But the truth is—some is, some isn't. Your body's metabolism is set to keep your weight and energy at a reasonable equilibrium. Eat less and the metabolism dips down, eat more and it speeds up to process the intake. Understanding what your body wants makes it easier to get what you want too.

According to the CDC, about 42 percent of Americans have obesity—a dramatic increase from the 13 percent in the early 1960s. The average weight of an American woman has changed from 140 pounds then to about 170 pounds now, while men have gone from 166 pounds to 200. However politically correct it has become to say that all weight issues are genetically based, evolution just doesn't happen that quickly. Something else is going on.

Once when I was the editor-in-chief of a national magazine, I had a well-known investigative reporter do an article about how the agricultural corn interests along with the commercial food industry had changed the diets of Americans—making us fatter and less healthy. The article was smart and fairly devastating. The CEO wouldn't let the piece run.

"Who do you think our advertisers are?" he asked.

Whether it's from overly processed foods or corn syrup

being dumped in everything we eat, many people aren't happy about how their bodies have morphed. About forty-five million Americans try a weight-loss diet every year despite the data showing that most diet programs don't work and yo-yo weight loss is physically and emotionally damaging. When you lose weight, you generally drop some muscle but typically gain back mostly fat. The muscle in your body is more metabolically active, so after a few yo-yo dieting experiences, you may weigh the same as where you started but burn fewer calories at rest. This is not a result that leads to happiness.

Instead of taking a different approach, diet companies change the language rather than the plan. The same messages get repackaged to current tropes. Weight Watchers is now simply WW, with the tagline "Wellness That Works," and new online favorite Noom promotes itself as encouraging "a healthier you." Both programs and most others end up counting calories. The new prevalence of weight-loss drugs like Wegovy and Ozempic puts the emphasis right back on losing pounds.

The paleo diet became widely popular by supposedly teaching the food plan that humans were originally created to eat. As far as diet cons go, this one wins for sheer audacity. The Paleolithic era lasted for some three million years, beginning from the earliest known use of stone tools to the time fifty thousand or so years ago when humans first began producing works of art like jewelry and cave paintings. Existing in many different climates, the Paleo people lived in a wide range of areas—from what is now Europe, Japan, and Australia to the Americas and the Arctic Circle. Does anyone really think that people ate the same thing for all that time and in all those places?

I suppose it's nice to imagine what our food choices would look like if we existed in a different time and didn't have candy bars flashing at us at every checkout counter. But we do have those temptations, and they create a whole new complication in trying to understand the messages from our bodies. The challenge becomes similar to the one that data scientists encounter—how to separate the signal from the noise. The signal is the information that matters. The noise is all the other distractions. In this case, the signal is what our bodies want. The noise is the candy bars. To make yourself happier, you need to be able to tease them apart.

The most logical approach is to stop following someone else's diet rules and start listening to your own body's cues. Instead of being on a diet or off a diet, think of eating as something you do every day to make yourself happy and healthy. Some nutritionists and dietitians now talk about "instinctive eating" where you give up weighing food, assigning points, obsessively logging what you eat, and considering some foods forbidden. Instead, you start paying attention to what you want and when you want it.

I call it the Happy Body Lifestyle. If you wake up hungry, you should eat then. If you're not hungry in the morning, have something very light and ignore the trope that breakfast is the

The Happy Body Food Plan

most important meal of the day. For you, it's obviously not. I'm often eager to eat at midday, so I'll eat a large lunch and afternoon snack, and I don't want much by dinnertime. My husband eats almost nothing during the day and is hungry for a big dinner at night. Though we are highly compatible in most things in life, our food timings are completely at odds. It doesn't have to be a problem, though. I've learned that I can enjoy our conversation at the dinner table while he has lamb chops and potato and peas and I munch on a salad and fruit.

The only rule on the Happy Body Lifestyle is to eat as much fresh, real food as possible rather than the Frankenfood that the food industry produces. If the ingredients list has a lot of words you can't pronounce, your body might be equally baffled. Your body knows what's going on when you eat an apple or a chicken sandwich on whole wheat bread, and your gut-mind connection makes a good assessment of how many calories you've taken in and how hungry you still are. But overly processed foods don't register normally with your body's hunger signals. Your mind can't accurately assess what you're eating, and as we saw with the fake milkshake study, your hormones can get led astray.

Researchers refer to foods as ultra-processed when they contain ingredients like high-fructose corn syrup, protein isolates, hydrogenated oils, and various chemical additives. Since 70 percent of packaged foods in this country qualify as ultra-processed, it may be whistling into the wind to warn you against them. But since we're talking about getting happier, you might be surprised to discover that the packaged cookies, breakfast cereals, chicken nuggets, and frozen pizza you eat for convenience could be undercutting your positive mood. One study found that people who ate large quantities of ultra-processed foods also reported more mild depression and anxious days. Right now there are several theories on why this happens, but no definitive answer. It may be because the heavily processed foods change your gut bacteria, or that they increase inflammation. One way or another (and possibly in many ways) your body isn't being fed right and your mind suffers.

Knowing this connection makes you think about certain foods in a completely new way. Next time someone offers you a doughnut, you might politely decline on happiness grounds. "No, I'm in a good mood. I don't want to ruin it."

Feed Your Body What It Really Wants

Your body is generally good at knowing when it needs vitamins, minerals, and protein. If you're eating the empty calories that the food industry churns out, your body is constantly seeking more nourishment. Your body can't tell you that it's full because that extra-large bag of potato chips doesn't have any of the elements your body actually needs. Even foods that promise to be healthy can backfire. A bag of "veggie straws" sounds like a good idea, but according to the Cleveland Clinic, they're heavily processed, lack fiber and protein, and are practically devoid of nutrients. You'd be much better off with some carrots dipped in hummus. Similarly, the Cleveland dietitians say that rice cakes "are really just a carb with little to no nutrition." Eat a lot of them and you feel virtuous (since they taste dry and unpleasant) but completely unsatisfied. Your mind is pleased that you've eaten something decent that should make you feel full, but your body doesn't have any of the actual sustenance it needs.

"Feed your body what it needs, and it will tell you when it's had enough," says Ruth Reichl, the renowned food writer, restaurant critic, and former editor of *Gourmet* magazine. "If you're eating food with very few nutrients, that time will never come."

Reichl says she became overweight as a teenager and gained even more pounds when she got to college. Trying various extreme diets, she felt miserable, starved, and unhappy—and the weight always came back anyway. Her life changed when she met and married a man who liked her larger size. Feeling comfortably loved for who she was, she stopped worrying about how skinny she might become and instead began focusing on what she actually *wanted* to eat. She paid attention to her body's hunger signals—and realized she had never actually done that before. Letting herself slow down and indulge in the sensory pleasures of cooking—chopping and peeling and enjoying the wonderful smells—she described feeling as if she'd consumed a meal before she ever sat down at the table. With this new attitude, her body started to change. She was stunned to discover she had lost thirty-five pounds.

I've had very similar experiences. One thing I have noticed over the years is that almost every time I decide I will go on a weight-loss diet, the result is that I gain weight. I begin thinking about food in a different way—what I "should" and "shouldn't" eat rather than just paying attention to my own hunger and eating what I want. I've never been particularly heavy, but in a pattern common for many of us, I have gained and lost the same ten or twenty pounds many times over. During the early months of the covid pandemic, I ate more than usual—notably large hunks of cheese—and gained weight. I didn't mind. I felt grateful to be healthy when so many others suffered, and life was too short to worry about a scale. The body positivity movement had given me that reassuring perspective.

Once I was back out into the world, I started swimming regularly, and as the activity made me more comfortable in my body, I also felt more attuned to its signals (a key advantage of exercise). I started yearning for healthy foods again and began to listen to my genuine hunger rather than my mind's desire. I didn't deny myself anything but I realized I was happier eating smaller portions. Junk food (and cheese) started to have less appeal, and I ate a lot of fruits and vegetables and fish and chicken. When the next summer rolled around, I realized that I had lost ten pounds without ever thinking about it.

That's the secret of the Happy Body Lifestyle. You think about what your body actually wants and needs. You focus on real food, but the occasional indulgence isn't a problem. It helps you return the word "diet" to its original meaning of the selection of foodstuffs we eat rather than its newer meaning of the restrictions we make and the food we *don't* eat. Accepting this can be a breakthrough. As a teenager, I read a diet tip in a magazine that I followed for years. It advised that if you had a craving for a frosted cupcake (or something equally indulgent) you should first eat some celery, then have a few carrots, next an apple, a yogurt, a slice of toast…and so on up the calorie pyramid. The idea was that you would be full before you got to the cupcake.

Now I've learned a different approach. If I want a cupcake, I'll get a small one and sit down and savor it as a treat. I'll enjoy it as a pleasure, not a failure.

Comfort Foods

One worry friends of mine have raised about the Happy Body Lifestyle is that without being given firm guidelines, they will overindulge. During times of stress, won't we rush to comfort foods like mac and cheese, pizza, or a juicy hamburger on a bun? The cliché in romantic movies is that the woman upset about losing a guy (he always comes back) tucks into a pint of ice cream for consolation. But don't worry about the clichés. The truth is that comfort food is more a media and marketing invention than an actual physical phenomenon.

People do yearn for different foods when stressed, but that seems to have more to do with mind than body. Every Valentine's Day, we hear that chocolate contains chemicals associated with happiness and love. In one large-scale study out of University College London, people who ate dark chocolate reported less than half as many depressive symptoms as those who didn't have it. An amazing result from a Hershey bar! But here's what's even more interesting. Your body metabolizes chocolate quickly enough that the supposed feel-good chemicals are gone from your system long before they can reach the brain.

Many people crave sweets as comfort food—a frosted cupcake or squishy brownie—and at least one study shows that under stressful conditions, sweets become more appealing. You could interpret that as an evolutionary response, a throwback to a time when we might have needed extra physical energy to get through a stressful time. But other comfort foods like pizza and pasta aren't particularly sweet, and it's unlikely that greasy french fries (another favorite) give that energy boost. While sweet foods have been linked to the release of mood-improving chemicals like serotonin and opiates in some people, don't get too excited by headlines claiming that ice cream and cookies can make you happy. The number of hormones released is not significant enough to truly change a mood.

It seems much more likely that a comfort food craving is simply your mind's attempt to use the sensory input of food to invoke a past memory or experience. Even if you've never read Marcel Proust's great twentieth-century novel *Remembrance of Things Past*, you've probably heard about the taste-memory that launches the story. The writer, slightly depressed and weary from a dull day, soaks a morsel of "plump little cakes called 'petites madeleines'" in a spoonful of tea, and suddenly a shudder runs through his body.

"An exquisite pleasure had invaded my senses…and at once the vicissitudes of life had become indifferent to me, its disasters innocuous, its brevity illusory."

As far as stress reduction, you don't get much more overwhelming than that. One taste of cake and tea and he has a sensation rushing through him as powerful (he says) as love. Trying to understand the source of his sudden prodigious sense of joy, he realizes it is connected to the tastes he has just experienced "but that it infinitely transcended those savours…"

Nobody would suggest that we need to figure out what neurotransmitters got released in Proust's madeleine moment. Food connects us to memories and, even more, to how we once felt. Our brains can be lazy sometimes when left on their own and need a prod to remember a happy sensation. The foods you ate during celebrations as a child or with your best friends during college can bring back the flood of feelings that you had then.

If you want to listen to your body to tell you what to eat, you have to be aware of how your mind gets involved in the action too. The good news is that letting food be a source of pleasure rather than guilt can make you more joyous in both mind and body.

But an interesting question occurred to me about that *New Yorker* cartoon where one person is eating chips while the other exercises. Which activity will really make you happier? If we put the question to a vote, I expect the chips might win. The evidence showing that exercise is a mood booster is extremely strong, but the thought of going to a gym or tying on our running shoes often leaves us conflicted. Finding out why it's so difficult to get ourselves moving turned out to be both more complicated and exciting than I would have imagined.

10

How Exercise Makes You Happy

> Life is like riding a bicycle. To keep your balance you must keep moving.
>
> —ALBERT EINSTEIN

SINCE ONE IMPORTANT ROLE FOR your brain is assessing what is going on with your body, slouching alone on the couch or sitting motionless at your desk is unlikely to make you feel joyful. On the other hand, when your body is moving with vigor and feels active and strong, your brain gets the message that all is going forward as it should. The brain processes the positive input and determines that you're happily on the right track. It can relax! No need to send out any stress alerts or alarms.

Sometimes it seems that your body yearns to exercise but your mind can't get up the motivation—while the reverse may also be true. You know you'll feel better if you get moving, but your body seems happy to mimic the position of a three-toed sloth. Something about this struck me as very odd. Our bodies are smart and they want us to stay healthy and alive. We seem naturally wired to let our bodies do the right thing. So…if physical activity makes us feel good, why do we find it so hard to get motivated? And how can we change that?

Doctors, anthropologists, and evolutionary biologists regularly say that our bodies were made to move, and for most of human history, we have walked and foraged and done physical work for a large part of the day. As recently as 1900, most people in the United States lived on farms or in rural areas and, according to the U.S. Census, "performed chores by hand, plowed with a walking plow, forked hay, milked by hand…" Fifty years later, the scene dramatically changed as electricity and power equipment took over many chores, and in the cities, office workers began pulling chairs up to their desks, seemingly never to get up again.

It's only recently that "work" for many people has turned sedentary while leisure activity involves sitting on a couch to stream TV, movies, and video games. What happens when people are suddenly sitting 13–15 hours a day? One study found a connection between prolonged sitting and a risk of early death, and other researchers report links between sitting and a metabolic syndrome that includes obesity, high blood pressure, and high blood sugar. Interestingly, the problems seem to lessen or even disappear if you simply get up more often. Most of the research doesn't particularly support the widespread media hysteria (*Sitting is the New Smoking!*) or the sudden market for standing desks. As one Harvard Medical

School doctor pointed out, sitting uses 80 calories per hour while standing raises that to 88—so if you stand for three hours, you can feel comfortable eating one extra carrot.

Given that movement is a simple way of making yourself happy, you'd think people would be pounding down the doors of gyms and rushing onto the local soccer fields at all times of day and night. But despite all we know about how movement activates the pleasure circuits of the brain, most people don't associate "exercise" with "happy," I understand. In our current culture, we've somehow turned the natural pleasures of physical activity into one more chore to undertake with gritted teeth.

Rather than walking through the woods or meandering through a pretty park where all our senses are aroused, we wake up early enough to schlep to the gym and get on a treadmill—which when you think of it is a rather bizarre device. Daniel Lieberman, a chaired professor at Harvard who studies the evolution of the human body, describes treadmills as noisy, expensive machines that make you work for no purpose and move without getting anywhere. He cheerfully points out that treadmills were used in Victorian prisons to keep prisoners from relaxing or enjoying themselves. Now they get your muscles moving but leave your brain unstimulated and disinterested. You experience no sensory changes—no fluttering leaves or chirping birds, no irregular ground, or unexpected views—and unless you're particularly fond of the smells, sounds, and sights in the gym, you probably put on headphones and try to distract yourself from the boredom.

Managing to get on the treadmill at the gym means you are a step ahead of most people. Despite their best intentions, about half of the people who get a gym membership drop it after six months, and even while paying the monthly fee, nobody shows up very often. One popular fitness chain signs on an average of 6,500 members per location even though their gyms have space for a max of 300 people. The business model relies on the fact that you'll feel guilty enough to join—but not guilty enough to get your money's worth.

Exercise shouldn't present such a conundrum. Your body is ready to make you happier and improve your mood if you would only do a little physical activity. Since your brain knows that exercise is good for you and will keep you healthy, your conscious mind has every intention of getting you on that treadmill. Mind and body seem to be joyously synced. But when the moment comes for action, the whole thing falls apart.

Why Your Body Doesn't (Always) Want to Exercise

Let's go back for a moment to the idea that our bodies were made to move, not to sit around idling. As we've seen, physical activity used to be a part of daily activities—necessary to find food, build shelter, raise children. Professor Lieberman points out that our hunter-gatherer ancestors often walked long distances each day and might have engaged in other activities like dance or sports for the pleasure of community. Physical movement was woven into everyday life. But here's the important point he makes: *They never exercised just to use up calories.*

Your body wants to preserve as much energy as it can for its main functions, including reproduction. The big human brain consumes a lot of energy, as do all of our other functions from heart to lungs to gut. A typical body requires 1,300–1,600 calories just to exist, and getting that many calories used to be hard work. You expended energy walking or hunting or foraging. Your wise body learned to preserve energy, not waste it on activities that won't lead to food or survival.

Even though you can now meet your daily calorie requirement with a single chomp of a Wendy's Big Bacon Cheddar Cheeseburger, your body hasn't caught up. It doesn't know about the drive-through Dunkin' Donuts and still wants to save calories. When you feel no urge to get off a comfy couch in order to sweat on a treadmill or work out on weight machines you're not being lazy or slothful—you're just responding to long-developed instinct. As Daniel Lieberman puts it, "No sensible adult hunter-gatherer wastes five hundred calories running five miles just for kicks."

Professor Lieberman has spent time in the woodlands of Tanzania with one of the last remaining hunter-gatherer tribes and also visited subsistence farmers and other indigenous groups largely uninfluenced by our technology-focused society. He believes the idea that humans should *want* to exercise is a myth. People in these traditional cultures work hard, he says, but they are also happy to sit or squat and relax around a campfire when there isn't an immediate need for activity. They would find the idea of training for an Ironman Triathlon completely preposterous.

"Whether you are a human, ape, dog, or jellyfish, natural selection will select against activities that waste energy at a cost to reproductive success," he says. "In this regard, all animals should be as lazy as possible."

This instinct to preserve energy shows up in unexpected ways. I enjoy taking long walks when I'm at our house in Connecticut and feel invigorated as I stroll alongside the river or saunter briskly through the town. But when I later drive over to the grocery store and turn into the parking lot, I reflexively search for a space as close to the main door as possible. If the nearest row of spots is filled, I feel a slight curl of irritation. I'll have to walk an extra twenty steps! It's not just me. Check out the parking lot in front of any gym or yoga studio and you'll see all the cars clustered near the entrance. It truly makes no sense at all. Everyone has come here to exercise, so why not walk a little farther and spend a little less time at the gym? It's an unconscious reaction—and perhaps reveals our deeply-wired inclination to save our energy whenever possible.

Daniel Lieberman tells the funny story of attending the annual meeting of the American College of Sports Medicine, an organization dedicated to promoting the value of exercise. The professionals who attend the meeting consider exercise to be crucial for health and well-being, and they are there to hear more about how to spread the gospel of exercise. As an informal experiment, Lieberman placed himself near a staircase and watched how many people took the stairs versus the escalator next to it. In the ten minutes he watched, he observed 151 people—and just 11 took the stairs. "Apparently, people who study and promote exercise are no different from the rest of us," he said.

People who live in big cities like New York or Chicago may walk a lot in the natural course of the day (taxis are expensive and subways are noisy, so it's sometimes the easiest way to get around), but in the rest of the country, walking is looked at with some suspicion. On a book tour to Houston a few years ago, I got a ride from the airport to the small hotel where I was staying and asked at the front desk about getting lunch. The helpful clerk handed me a list of several restaurants and said the first one was the closest, about ten minutes away.

"A ten-minute walk?" I asked.

"No, that's driving time!" he said with a laugh. "There's no place at all that you can walk!" He gestured to the wide busy highways that crisscrossed outside, and when I suggested that there must be sidewalks somewhere, he looked alarmed that I might actually venture out and picked up a phone. "Let me find someone who can drive you," he said anxiously.

For most of human history, people walked several miles a day to get food and water, but we have reached the point now where the idea of walking to lunch can make a kindly clerk think you must be slightly daft.

How to Love Physical Activity

So here we have the dilemma. The reward systems associated with physical activity, including the release of dopamine and other neurotransmitters that make you feel good, may be evolutionarily tied to our previous *need* to move. Since physical activity was always in the service of getting food and staying safe, it mattered that you got positive reinforcement for your efforts. Your body also learned to preserve energy when there wasn't an immediate need for it to rev up. Now we have very little requirement to move for daily living or survival, and the "preserve energy" instinct remains—which makes us prime candidates for not moving much at all. But we still need physical activity in order to get an all-important shot of positivity and mood-boosting endorphins.

Now that physical activity is a choice, how do you get yourself inspired to take the first step? You could try the rational approach. Experts have very little doubt about the power of exercise to improve your health, and we've all read plenty of articles on how exercise will help stave off cardiovascular disease, diabetes, and some kinds of cancer as well as improve sleep, lower stress, and possibly slow cognitive decline too. But describing the advantages that way puts exercise in the same category as brussels sprouts and kale salad—good for you and pleasant if done right, but not necessarily your first choice in picking from life's buffet.

You're most likely to stay with an exercise plan if you find something your body actually loves to do. Frame exercise as joy that you are allowing yourself and everything starts to change. Try to think back to the activities you naturally did as a child. Were you the person who loved to go to ballet class or the one who ran joyously down the street and scrambled up trees? It could give you some insight as to whether you'll be excited going to a barre class now or signing up at a rock-climbing gym.

I liked swimming as a child, and my happiest memories from summer camp involve hours spent in a very murky lake. (Being bitten once by an unidentified creature and rushed to the local doctor didn't dissuade me.) As a young adult, I spent many hot weekends swimming laps in a big community pool, but after that, swimming became a very occasional activity. I'm sorry to admit that I passed up many swim opportunities over the years because I didn't want to put on a bathing suit.

Recently, I decided to put aside silly concerns and give swimming another go. I signed up for thirty-minute sessions of lap swimming at a small local pool—and the first time I got in the water, I felt like I was giving myself an amazing treat. The thrill has remained. I now catalog swimming in my mind as pure pleasure. When I glance at the overhead clock during a swim, I'm invariably disappointed to see how little time I have left. "Only ten minutes more? I want to keep going!" It's the exact opposite feeling I have when I'm on an elliptical machine or stationary bike and can't wait for the grueling session to end.

Exercise should be a joy, not a chore. If we were driven only by rationality and reasoning, knowing that a particular exercise could improve health and longevity would be motivation enough. But it's not. Our bodies and minds like being happy right now. To keep yourself inspired, you need to find the exercise that feels like a sensual indulgence rather than a punishment. That way you'll gain the immediate advantages of sensory satisfaction—and the longer-term payoffs of health and energy.

The simple solution is this: If you want to be happier, make sure you have some movement in your life every day. Walk your dog or throw a Frisbee with a friend. Allow yourself the pleasure of strolling through a local park or along a waterfront or just in your own neighborhood. There are many answers to the happiness-health-exercise equation—and you should give yourself the pleasure of finding your own. Remember that your body wants to move—and it will reward you for any activity you do with improved mood and attitude.

Researchers love to study exercise because they can put people on treadmills or elliptical machines and wire them up to measure heart rate, blood pressure, and lung capacity, then take blood samples to discover what hormones, chemicals, and neurotransmitters are circulating. Their published results offer incredibly granular insights on the best time of day to exercise, the best intensity, whether short bursts of exercise are as good as long duration…and on and on. Most studies I've read are carefully done, and I don't question the accuracy or integrity of the data even when the findings can be contradictory. (Should you do low intensity for a sustained period of time or intervals of high intensity? I've seen findings for both.) The bigger point is that unless you're already a high-level athlete, none of it will make a difference in your life. This is the case where you shouldn't let the perfect be the enemy of the good. Let the researchers argue whether six thousand or seven thousand steps is the ideal—all you have to know is that movement of any kind will increase your feelings of happiness, joy, and gratitude.

Exercise and the Feel-Good Chemicals

Physical activity causes the release of various endorphins and neurotransmitters that interact with brain receptors, creating feelings of happiness and pleasure. These naturally produced chemicals are as powerful as many drugs. Evolutionary biologists agree that the happiness-making hormones in our bodies were once deeply important for our survival—and maybe still are. In the days when people were out on the plains for hours hunting animals and foraging for food, they needed more than determination to keep themselves going. Having natural chemicals in the body that produced a sense of well-being after exertion would have helped them persist. It took until the early 1970s for researchers to come upon endorphins which circulate in the body and fit the evolutionary bill. Candace Pert, one of the key scientists in the discovery, described them as "the body's own pain suppressors and ecstasy inducers." Even if you now get your dinner delivered from Grubhub or Uber Eats, endorphins will kick in to make you feel good after any vigorous activity from dancing to judo to swimming.

Another feel-good chemical was identified in 1992 by researchers who first discovered a receptor in the brain for THC, the active ingredient in cannabis. Our bodies are very smart (as we are discovering), and they wouldn't waste space on a receptor for something we couldn't produce ourselves. (Evolution was not counting on weed being legalized.) Researchers soon found anandamide, which is very similar to THC and has a similar effect in making you calm and happy.

Long-distance runners have made us aware of the "runner's high"—a sense of euphoria they reach after a certain number of miles and exertion. For many years, it was attributed to endorphins—but eventually some scientists questioned whether the endorphins circulating in the body actually reached the brain. The discovery of anandamide brought along a whole new twist. Like endorphins, it is released during vigorous exercise. Arne Dietrich, a fascinating guy who does neuroscience at the American University of Berlin, thinks it may be the secret to the exhilaration that some people experience with exercise. "Anandamide is a tiny little fatty acid that crosses the blood-brain barrier like nobody's business," says Dietrich cheerfully.

Some combination of endorphins, dopamine, and anandamide that are revved up with movement improves your mood in the moment and for hours or even days after. "When

I can't play squash for a couple of days, I get really tense and grumpy," one middle-aged guy told me. Even casual athletes who get injured can find themselves getting more depressed and anxious than usual. You don't actually get addicted to exercise (except in very rare cases), but the change in body chemistry can show up in your mood.

My friend Leslie gets up at 5:00 a.m. most mornings so she can exercise before she begins her very full day working as a jewelry designer. Most weekends, she and her husband drive an hour or two to hike on mountain trails, and they consider exercise to be nonnegotiable. "We need it to stay happy," she told me. I don't know if Leslie and her husband would be less fun to be around if they didn't exercise, but I expect their high spirits wouldn't be quite as high.

Get a Gardening Buzz

You definitely don't have to become a long-distance runner to get the advantages of movement—almost any kind of physical activity can give you a positive emotional buzz. Going for a walk, tending your garden, dancing around your living room, or playing tag with your toddler all change your body chemistry in a positive direction.

The emotional changes that exercise brings are as important as the health effects. One large survey of people in twenty-four countries found that moderate physical activity led to higher life satisfaction and happiness. The exact level of activity that improves mood and happiness is hard to pin down and may vary for different people. Some studies say that small bursts of exercise are enough to improve your mood and others show that happiness increases with the more activity that you do. A major study out of Taiwan looked at young, middle-aged, and older adults (all basically healthy and mobile) and found that at all ages, life satisfaction and happiness improved as people became more active. Regardless of other factors that also affect happiness, like marital status, education level, and income, being highly or moderately active consistently led people to have a more positive feeling about life in general. Satisfaction was lowest for those in the "low active" groups.

The lone dissenting voice to the more-is-better theory that I could find came from a study in Germany showing that moderate-intensity exercise has a "significant and positive effect" on well-being, but very vigorous activity can actually send your mood in the opposite direction. Nothing else I could find supported that—and I certainly know a lot of happy marathoners—but for now, I'd say that, while exercise definitely works to make you happier, the amount and kind are still up to you.

How Exercise Cheers You Up

Physical movement can change your mood from good to better, and it can make a lousy day suddenly feel a little brighter. If your mood has sunk even lower, exercise will combat despair and help ease symptoms of depression. It works for people of all ages and with symptoms ranging from mild to severe—and the evidence is so strong and far-reaching that it's hard to dispute. Walking as little as thirty minutes a day for a couple of weeks can produce a significant reduction in depression, and exercise seems to *prevent* depression too. Stay active for just a few hours a week and your risk of depression goes way down. Physical activity improves well-being for people with and without clinical symptoms.

In research comparing exercise to antidepressant medications, the results are close to a draw. One study divided depressed men and women into three groups with three different treatment plans—one group got straight exercise, another received antidepressant medication, and the third got exercise plus medication. The medication worked the quickest. But before you rush out to fill or renew a prescription, think about this. After sixteen weeks, there were *no* significant differences among the groups, and at the ten-month follow-up, those in the exercise group had dramatically lower rates of depression than any of the others.

Finding this clear connection between exercise and depression relief has been easier than figuring out why the link exists in the first place. I like to think of it as your body looking out for your well-being. If you're depressed, your physical health also tends to deteriorate, and it's your body's job to try to keep you well. Exercise increases the levels of neurotransmitters in your blood and probably in your brain—which leads to positive mood improvement. When you're depressed, you may not have enough serotonin, dopamine, and norepinephrine circulating in the brain, and it appears that simple movement can change the imbalance.

Your body also vibrates with a new energy when you're moving and active, and it sends positive signals to your brain that you can perform whatever needs to be done. Psychologists say that depressed people often feel like they can't exert any control over their environment or circumstances. They want life to be better, but they feel helpless and powerless to get to a better spot. When your body is moving and active, it sends a different message to your mind that may counter some of the negativity. Your body is coping and sending clear messages of what psychologists call self-efficacy—or the ability to control your own behavior and environment. Whether you're swimming, running, doing yoga, or dancing, your body feels in control, and it lets your mind know that things are okay.

You don't have to be a great athlete to get the advantages of self-efficacy—your body is happy to reward you at any level at all. When my husband comes back from a run (his favorite activity), he has an extra swagger about him, and though he regrets that he's not as fast as in his marathon-training days, running outside gives him a strong emotional boost. His body sends his mind the signal that he remains strong and competent. I get a similar feeling of confidence when I'm swimming. Moving through the water, I feel sleek and powerful, and even when (as happens often) a much faster swimmer in another lane makes it very clear to my conscious mind that I'm not particularly powerful at all, my body refuses to agree. It's moving exactly the way it should! It's accomplishing the task at hand!

Your body wisely produces feel-good rewards when you are active to keep you positive and moving forward. It does not want you to give up and opt out. Given the slightest chance to move, your body will work hard to make you happier every day.

The Best Reason to Exercise Isn't the One You Think

The driving force for a huge percentage of people who join gyms or buy at-home exercise equipment is a desire to lose weight. At the beginning it seems to make sense. Exercise burns calories, and even if it's not a lot—say two hundred or three hundred calories for a workout—the deficit should eventually build up and help you drop pounds. As an added advantage, the "calories used" monitor flashing on your treadmill or stationary bike tells just part of the story. Depending on how vigorously you're working out, your resting metabolic rate can stay elevated for several hours after you finish. Researchers refer to this as post-exercise oxygen consumption, but everyone else calls it afterburn. It seems like an unexpected gift. You've left the gym, but your body remains active.

All of this sounds encouraging, but remember that it's very hard to trick your smart body. It wants you to have enough energy to survive, so if you use up calories exercising, you may find yourself a little hungrier and eating an extra snack or two. Your body has a back-up plan, so if you don't eat the extra calories, it may slow down your metabolism enough to keep your overall energy expenditure constant. You may be making a big sacrifice to get to the gym (waking up earlier, putting on your exercise clothes, driving across town, doing your workout, taking a shower…), but your body needs to make only a small tweak to make it calorically meaningless.

Here's how it works. Your resting metabolic rate accounts for about 70 percent of the energy your body uses in a day, and simply digesting food accounts for another 15 percent—for a total of 85 percent spent on basic living. That leaves only about 15 percent or slightly more for all the physical activity you choose to do. When you increase that small part of energy expenditure under your voluntary control by exercising, your body makes sure that your overall energy output doesn't change much. It makes small metabolic shifts to adjust the 85 percent of energy that keeps all your systems functioning. Many evolutionary biologists believe that those long-ago hunter-gatherer ancestors who walked all day and dug for food expended about the same total energy as we do now sitting at our desks and staring at our computers.

I mention all this not to discourage you from exercising but to say you'll be much better motivated in the long run if you're exercising for reasons that actually work. Researchers continue to debate the finer points of what small effect exercise might have on weight control, and it seems it may play some role in maintaining weight loss. But you'll be much better off if you forget the scale as a reason to exercise and focus on the areas where we know exercise makes a difference—including your happiness right this very moment.

Much of what I'd been thinking about until now had to do with how our bodies can be joy enhancers, but I had to admit that there are times when our bodies are joy deniers instead. Everything seems to be going along swimmingly, but then you stub your toe or break your ankle, your back hurts or your shoulder aches, you get stomach pains or a pounding headache. You might gobble down aspirin or other meds to cure the problem, but sometimes the causes of pain are more elusive than we realize. My next goal was to find out what our bodies know about pain that we should understand too.

PART FOUR

How Your Brain Resolves Pain

Our bodies can inspire positive feelings and emotions, but a big question emerges on the other side. Why does your body sometimes make you miserable rather than happy? Read on for a completely new perspective that can forever change how you think about pain.

Everybody Hurts (Sometimes)

> The great art of life is sensation, to feel we exist, even in pain.
>
> —LORD BYRON

WE GENERALLY THINK OF PAIN as being a straightforward physical problem. Whether you have a pain in your neck, stomach, foot, or left pinky, the first thing you want to know is *What's wrong?* We expect our bodies to function in predictable ways. Break a bone in your arm, and it will hurt, and when the bone mends, the pain will go away. If you have a migraine headache or a stiff neck, you want to find the cause of the problem so it can be cured. But pain turns out to be incredibly complicated—an interweaving of mind and body that's not easy to unravel.

When I started having some back aches a few years ago, I knew I wasn't alone. Everyone's back hurts. Nearly sixty-five million Americans say they've had a recent episode of back pain, and it's been reported that 80 percent of people will suffer from back pain during their lives. While that may be an exaggeration, the odds are enormously high that you'll have a back problem at some point. Most back pain goes away in a few weeks with Advil and mild exercise, but once the pain starts, panic often sets in. *My back hurts! I can't bend over! I'll be this way forever!*

I had the same problem. The more I thought about my back, the more it seemed to hurt. I looked up treatments for back pain, and the long list from the Mayo Clinic included opioids, steroid injections, muscle relaxants, implanted nerve stimulators, and surgery. Yikes, no thank you. What was causing the problem? I had no idea—and no particular way to find out. Consult a doctor about your back pain and there's a 50 percent chance that they'll send you for an X-ray or MRI scan. This might make you happy that something is being done—but don't be fooled. Research shows that except in the most severe cases, those tests are completely useless. Spinal "abnormalities" like a bulging or herniated disc show up on both people who have pain and those who don't. Conversely, some people with seriously bad findings on scans have no pain at all. In one famous study published in the highly respected *New England Journal of Medicine*, people who had no symptoms of back pain were given MRI scans. The radiologists who read the reports didn't know about the symptoms (or lack of them)—and they diagnosed two-thirds of the people as having disc problems.

Researchers at Oregon Health and Science University

added more support to the scans-don't-work literature when they reviewed clinical trials involving nearly two thousand patients and published their findings in the similarly respected science journal *Lancet*. Lead researcher Roger Chou explained that doing routine scans "does not lead to improved pain, function, or anxiety level, and there were even some trends toward worse outcomes." Those worse outcomes have several causes. If you're told there's a functional problem with your back, you may become tentative about moving around too much and completely stop exercising. It's counterproductive because movement is one of the best ways to get better. You're also more likely to end up with higher anxiety and tension—which are the *worst* ways to get better.

Since physical findings often don't correlate with how you feel, I decided to try some simple fixes. I looked for a better position for sleeping and bought a new desk chair. When lugging my computer to work made me feel worse, I bought a different bag and then a different computer. Nothing helped. The feeling of helplessness—that my body was out of my control—was even more upsetting than the pain.

I finally mentioned the problem to my husband, the doctor. He agreed that the X-ray route would not be helpful and suggested strengthening my abdominal muscles. When you have a stronger core, you don't put as much strain on your back. It was excellent advice, but I didn't take it all that well. I immediately began fretting that I'd gained weight and my husband thought my stomach looked saggy. (This is why doctors don't treat their own family members.) I finally got over the affront and tried to figure out what I could do.

I searched online for "core exercises for back pain," and I found a lot of them, including a seemingly endless supply of YouTube videos. For the next few days, I propped my iPad on the floor each morning and did an easy ten-minute workout with an online physical therapist who managed to be both sensible and encouraging. At night, if my back hurt when I lay in bed, I did one of the in-place abdominal-tightening exercises he recommended, and I felt instantly better. Perhaps the biggest change was that instead of feeling hopeless when I had a pang, I had a new sense of control. There was something I could do. I was working with my body rather than being afraid of it.

Was I convincing myself that I felt better, or was my back really improving? As we've seen already in so many other instances, body and brain aren't separate; they are constantly influencing each other, and that is certainly the case with pain. The neural circuits that cause pain travel from body to brain and brain to body and round and round. After I spoke with several pain experts, I started thinking about pain as being like the electricity kits my kids used to play with when young. The sets had wires and connectors and switches, and the idea was to put them together so they could make a small red bulb light up. Pain is like that red bulb. There are lots of places in that circuit where it's possible to intervene and stop or lessen the pain—and we need to find all the switches that we possibly can. For my back pain, the abdominal strengthening was one switch. But so was the comfort I got from feeling that I had some control over how I felt.

Your Personal Pain-O-Meter

The body's first response is always to protect itself—and your body uses acute pain as a warning sign. Step on a rock, touch a hot stove, get bitten by a snake, and you have an immediate reaction. Your sensory neurons send the information to your spinal column, which causes you to pull your hand away or jump back. In another split second, the information gets sent from other neurons in the spinal cord to the thalamus, the part of the brain that receives pain fibers. The thalamus sends signals back and forth with other brain regions, so it will get input on responding from other parts of the brain. Now is the moment when you shout "Ouch!" and your brain steps in to decide whether the snake looks poisonous or not.

All of that is well understood. What's a little hazier is why some pain doesn't go away. Most research suggests that the first part of the system—the nerves sending the emergency pain message toward the spinal cord—is incredibly efficient. You don't have to think about it, and acute pain gets an instantaneous response that doesn't yet involve your brain. Once the neurons to the brain get involved, though, things can go wrong. The nerves may keep sending the pain alarm even after crisis is over.

Dr. Sean Mackey, chief of the division of pain medicine at Stanford, says there's often not a direct link between an injury that happens to the body and the ways the brain responds. "Pain is an experience, a human experience, that's incredibly subjective," he told me. "It's no different than the experience of love, the experience of anxiety, or the experience of fear. The one distinction, and where we get confused, is that there is usually a physical stimulus we can connect to pain. We assume, therefore, there's a one-to-one relationship with the physical stimulus, but there's frequently not."

To help explain how that works, Mackey used the analogy of looking at a painting. The physical stimulus in this case occurs when the image hits the photoreceptors on the retina. The signals then pass to the brain, where the retinal image gets processed and modified and interpreted. At that point, one person may declare the image a masterpiece and another a piece of junk—*even though the retinal image is the same.* Something very similar happens when the body registers pain. The physical input doesn't mean much until the brain has processed and interpreted it, often adding meaning that isn't really there. The most important breakthroughs in pain management in recent years haven't been in changing the physical input or the site of the pain but rather in interfering with how the brain understands what's going on.

Smart, thoughtful, and attentive, Mackey is a much-admired researcher, but he also struck me as the deeply caring clinician you'd like to meet if you're experiencing pain. Though he's currently working to develop biomarkers for pain, he says that "we don't yet have a valid pain-o-meter where we can objectify pain. The individual experience belongs to you." He understands the goal is always to change how you perceive your own physical input—whether as masterpiece or junk.

When you first experience an injury, your body rushes a signal to the thalamus, which functions like Grand Central Station, sending the pain message off to other areas of the brain. Your body is extremely efficient in sending signals, but it's in the post-thalamus brain regions that the mess occurs. These higher brain regions are busy with many inputs and issues and can't possibly give equal attention to all of them. The pain signal becomes enmeshed with various other emotional matters. How much pain are you feeling? The pain-o-meter in the brain is now being influenced by all sorts of emotions, experiences, and expectations that really have nothing to do with the initial injury.

Why would emotional issues affect your experience of pain? Here's an interesting link. Neuroscientists have mapped a variety of emotional states directly onto specific areas of the brain, including the anterior cingulate cortex and the somatosensory cortex. The names sound complicated, but please don't worry about them. What's important to understand is this: The pain signals get sent *to these same regions*.

"Areas that shape your experience of pain overlap with areas of emotional and cognitive information," says Mackey. "The complexity is rather beautiful and elegant."

Your brain tries to filter the vast amount of input it

receives, creating a loop that amplifies some signals and turns down others. If you have significant anxiety or fear around the injury, or if you're tired or depressed or otherwise upset, the stress and frustration all amplify the pain.

Pain charts that now hang in hospitals and doctors' offices make it seem like pain can be given an objective number. They typically have a graphic with a green smiley face on one end (1 = no pain) going all the way to a red frowny face on the other (10 = worst pain possible). But your 6 may be my 3—and what does it all mean anyway? In many different experiments, people subjected to the same experience, like plunging their hand into ice water or having a hot pad placed on their arm, judge the pain dramatically differently. Some say it's not painful at all and some say it's excruciating—and everything in between.

The wide range of people's responses in these experiments tells us a lot. The hot pad that touches an arm is the same heat intensity for everyone, and the sensory neurons that register it are pretty accurate. The body has this under control. If the pad is hot enough, you will withdraw your arm instantly—a response from the spinal column that doesn't require the brain's involvement at all. But once the spinal column passes the signal on to the brain, the discrepancies in pain perception occur. Once various memories, emotions, and expectations are mixed into the pain circuit, the experience of what you're feeling has less and less to do with the actual physical event.

Okay, I'm going to pause here for a moment, because if you have back pain, neck aches, migraine headaches, chronic stomach distress, or any of a number of other chronic pains, I imagine that you're starting to get frustrated about now. You want to shout that *I'm in pain and it hurts and it's not in my head!*

Let me say…Dr. Mackey feels your pain, and so do I. Of course you're right, and of course it hurts. Nobody is suggesting that you're making it up. Recognizing that there is a component to pain that goes beyond the localized injury isn't meant to undermine the pain that you feel. Yes, you feel it! People often get defensive when told that pain involves their mind, and Mackey says he often explains that "Pain isn't in your head, but it is in your brain."

"Isn't your brain in your head?" I asked him.

He laughed. "It's a nuanced language issue. When people say 'It's in your head' they mean it's not real. You're imagining it. What I'm trying to say is that the experience you're having is very real, and it's a consequence of the brain circuits generating the experience of pain."

All Pain Is Real

Almost all of the pain researchers I encountered echoed Dr. Mackey's comment. They wanted to make it very clear that talking about mind-body loops does not mean that people are being blamed for their pain.

"All pain is real," one physical therapist told me.

"All pain is real," a neurologist insisted.

The mantra has become part of the discussions of mind-body

interactions. All pain is real. You feel it. You experience it. Whatever its source in brain or body, you're not making it up.

While I was thinking about what it means for pain to be "real," I came across some interesting research about taste done by the neuroscientist Charles Zuker, a Howard Hughes Medical Institute Investigator. His group showed that when taste receptors on the tongue send information to the brain, the sweet and bitter tastes go to specific cortical fields. (The cortex has the "little gray cells" made famous by fictional detective Hercule Poirot.) The sweet and bitter cells from the cortex then connect to specific regions in another part of the brain, the amygdala, which gets involved in emotions and makes value judgments on sensory input. *I like the chocolate cake! Don't like the onions!* That's about what you'd expect from a body-mind connection—and sounds a little bit like the pain circuit. The body sends the sensory information, and the brain registers it and interprets.

But now here comes the amazing part. In experiments with mice, the researchers found that by stimulating the correct neurons in the brain, they could make the mice behave as if they were experiencing specific tastes. The researchers gave the mice neutral-tasting water but stimulated the brain region responsible for the sweet circuit—and the mice drank eagerly. When the bitter circuit was stimulated, the mice lost interest in drinking. The researchers could even override the messages from the actual taste sensors. When the mice got a bitter quinine to drink but had the sweet circuit in the brain stimulated, they drank more than expected.

"Simply activate a few hundred cells in the bitter cortex and the animal not only thinks it's tasting bitter but executes all the associated behaviors," Zuker says. "The message from that was taste is really in your brain."

Taste is in your brain! Sure, you (and the mice) have taste receptors on your tongue and they are a key part of the process. But the connections between body and brain are thrillingly complex. We generally think of the five senses—taste, touch, sight, sound, and smell—as being fully physical functions, but once the body sends the message and the brain takes over, all sorts of twists and turns can occur. You drink something sweet, but if your brain decides it's bitter—well, bitter it is.

At the moment there are no practical uses for Zuker's research with sweet and bitter tastes, but it's pretty cool to imagine that someday you could eat a piece of melba toast and have your brain tell you that it's a gingerbread cookie.

You probably see where I'm going with this—because it would also be nice to stimulate the right little gray cells in the brain and make your back pain go away. To do that, you have to accept that no feeling, sense, or pain is all in your body or all in your head. It's both. All taste is real. All pain is real. Where it starts and ends is the bigger question.

Your Oversensitized Brain

Many researchers now think that chronic pain occurs when the brain gets overly sensitized to the pain circuits. Clifford Woolf, a neurobiologist at Harvard Medical School, explains that pain is not simply a measure of some "peripheral pathology"—such as a broken bone. It can also be the consequence of "abnormal amplification within the nervous system"—or what has come to be called central sensitization. The problem is no longer the broken bone but an overexcited nervous system that keeps firing neurons. The more it fires, the more ingrained it becomes and the harder it is to turn off.

I think of this as the lifting-weights theory of pain. Just like your muscles get bigger when you lift weights every day, your pain circuits get stronger and more sensitized as you use them. Eventually, you don't need to hold the weight to flex your bicep—just like you don't need the biomechanical problem to feel the pain. Most researchers now refer to pain as being a biopsychosocial problem, meaning that lots of different factors are coming into play. The biological or mechanical signal from your body is just part one. Emotional, cultural, and social expectations get involved too, and anxiety about the pain just makes it worse.

Sometimes the problems can overlap. If you have arthritis, the bone grinding on bone causes pain signals, and the continued action also leads to that central sensitization, so your pain has two sources. Fibromyalgia and similar diagnoses that describe unexplained pain throughout the body are probably examples of the central nervous system messing up. The nerve pathways become so oversensitized that they can't easily be turned off. Body, brain, mind, and emotion become completely interwoven in pain circuits. You can try to activate an off switch anywhere on the path.

Not all pain is the same. In addition to hearing the term "central sensitization," you'll sometimes hear about "neuropathic pain," which means the pain-sensing nerves are injured or damaged, as occurs when you suffer with a case of shingles. Similarly, when a nerve in the spinal column sends shooting pains down the leg, you are suffering from the one kind of nerve-related back pain that might be improved by surgical intervention.

The acute pain that occurs when you damage tissue, nerve, or bone is sometimes referred to as "peripheral nociceptive pain." It results from tissue damage or inflammation and is the sign of an immediate problem. The body has a great warning signal to prevent future injury. If you're running and twist your knee, your body alerts your brain that something is wrong—and the pain tells you to stop running. This kind of pain should resolve on its own. (Keeping the leg elevated and icing it now and then will speed the process.)

Pain becomes complicated, though, when this kind of acute nociceptive situation morphs into chronic pain. You assume that the problem remains the same—*That darn knee injury!* But interestingly, while acute pain is processed in one region of the brain, after a certain period of time the brain activity switches to different circuitry. Researchers generally describe chronic pain as anything lasting longer than three months. At that point, the pain circuit passes through brain areas that also process emotion, fear, and anxiety. At some point, your brain may start sending pain signals that your body doesn't even know about.

A very lovely woman I met told me she had been in a car accident a couple of years earlier and was rear-ended so forcefully that her car had to be towed away. The whiplash left her with a neck injury, and the pain had never gone away. She'd recently had an MRI and several other scans, and the doctor just gave her the good news that the damage had healed. But she didn't see it as good news. She remained in pain every day and had now made appointments with several other specialists. "I need someone to figure out what's wrong and where the damage is," she told me.

I thought of telling her a little about my research and the possibility that her body had healed but her brain circuits had not. But it was too complicated for a casual encounter—and probably explains why most doctors don't get involved in the discussion either. It's much easier for someone to want to pinpoint the physical problem that needs to be solved than to start considering the complications of other body-mind involvements. I bit my tongue because I didn't want to suggest in any way that she was making up the pain. I knew she hurt! As Dr. Mackey would have explained, the problem wasn't in her head—but it was (possibly) in her brain.

How Pain Gets Constructed in the Brain

If our body's pain signal is only a small part of our experience of pain, what are we supposed to do? To understand more, I got in touch with Tor Wager, a chaired professor of neuroscience at Dartmouth College who runs a lab that studies (among other things) how expectations and beliefs affect our brain and body. Like Dr. Mackey, he has been part of the big shift in how pain is understood.

"We used to think that pain is something that happens to you—a signal from the body that the brain just registers. But the whole field has shifted to thinking of pain as a signal that's *constructed* in the brain. Depending on what's happening in our environment, our brain can ramp pain up or ramp it down," he told me.

Wager's team did a study of 150 people with chronic back pain and randomized them into three groups. The participants had dealt with pain for an average of ten years and sometimes much longer, and all of them said pain interfered with their daily activities. The core intervention for one group involved eight sessions (two per week) with a therapist who helped them develop an alternative understanding about what pain means. Instead of attributing their pain to tissue damage or injury, they could consider that their pain was caused by sensitization in brain circuits.

"Pain is essentially a guess by the brain about what it should be feeling," said Wager. "If you learn that the pain is nothing dangerous and you don't have to be afraid of it, you can even laugh at it. You might even get to the point where you say *Wow, isn't that weird? My brain's playing tricks on me!*"

While people with pain are often told by their doctors to move carefully or not at all, this group was given the opposite advice. They learned to put their new theory about pain into practice. Even if the back hurts, it's okay to move into it. In fact, it's more than okay—moving is good and can make you feel better! You won't become disabled, and the back won't degenerate. The results were stunning. After the four weeks with a therapist, two-thirds of the people reported being pain-free or nearly pain-free. In a one-year follow-up, they continued to be doing well.

Flip the Pain to Joy

The lessons from Wager's intervention with a therapist can be helpful for all of us. Pain and the fear of pain cause people to move less and worry about simple activities from raking leaves to playing with a child. As you do less, the pain increases—because movement and activity will actually help with the blood flow. The more fearful you are of pain, the more it will hurt.

Dr. Mackey told me a story about a patient who consulted him a few years ago and came to his office on crutches. A master's tennis player, he had pain in his foot and had been told that if he put any weight on it, he would injure it further. Since he lived for sports and used physical activity as his stress outlet, he was thoroughly miserable. After examining him, Mackey diagnosed a Morton's neuroma—essentially a bundle of nerves in the foot that fire off when there's pressure. It's fairly common and very painful, but it's not dangerous.

"I told him, 'Listen, you're not causing any damage to your foot when you walk on it or play tennis. It's just causing pain.' He said, 'You mean playing tennis won't cause so much damage that I end up in a wheelchair?' I said, 'Of course not. It will hurt but you'll be fine.' He nodded and said, 'Okay, thanks. We're done here.' And then he just got up and walked out of the room without crutches."

I laughed. "You're like the healing waters at Lourdes," I told Dr. Mackey. "People visit you and throw away their crutches!"

He smiled. "He still had the pain. It just had a different meaning for him."

Helping people think about their pain in a different way can have huge effects. Pain comes from an individual sensory and emotional response that belongs specifically to you. The question becomes how to break the circuit. All the pain fibers eventually go to the brain—so you have the opportunity to turn off the pain at the local spot where it originates or once the signals have been sent along the chain. You can think back to those electricity kits again, with the various switches that can turn on and off the red light. As long as they're on, the electricity keeps flowing. As with pain, all it takes is one switch to turn off the whole system, and it doesn't matter where on the circuit it appears.

For treating chronic back pain, Dr. Mackey has had some success with an electric coil that he places on top of a patient's head. The "brain zapping" (it doesn't hurt) generates a focused magnetic field that influences the brain centers that register pain. "We've found that, with certain sequences of pulses, we can induce long-term changes in some people," he says. He and his team are still trying to figure out who is most likely to respond to the treatment.

I should add a caveat here that if you have an acute injury, you do want to rest until it's healed and recognize the pain as a danger signal. Similarly, if your back problem includes radiating nerve pain shooting down your leg, you need to deal with the physical issues. But millions of people have chronic back pain without being in either of those categories. Wager estimates that 80 percent to 90 percent of people with aching low backs don't have a structural cause for the pain. Even if the problem started with a localized peripheral cause—you shoveled heavy snow or lifted a squirming child—the difficulty usually comes after some months have gone by and you can no longer discriminate the physical pain from the brain circuit sensitization.

Your body uses acute pain as a warning that something is wrong, and the problem comes when your brain forgets to turn off the signal. The pain siren begins sounding all the time.

It's a little like an alarm going off in your house when someone opens the front door. You want to check it out, but whether you have an intruder or it was a false alarm, you also want to turn the alarm off. No good comes from having the siren go on and on and on.

The good news is that there are more ways to turn off the alarm than you may think. You just have to learn the right passwords. I was excited to start discovering some of them.

12

Pain, Pain, Go Away

> There is a thin line that separates laughter and pain, comedy and tragedy, humor and hurt.
>
> —ERMA BOMBECK

DURING THE 2012 OLYMPICS IN London, an athlete named Manteo Mitchell was running the first position on the 4 x 400 relay for the United States during a qualifying round. Watch a YouTube video of the race now and it doesn't seem like anything very dramatic occurs during his lap. Mitchell finishes just 1.5 seconds behind the leader and passes the baton smoothly to his teammate. Though one of the commentators mutters that Mitchell didn't seem to be running his best, nobody realized then that the performance was truly astonishing. Halfway through the lap, Mitchell broke his leg. His left fibula snapped, but he barely broke stride and continued at virtually the same pace for the next 200 meters. Later, he admitted that he'd felt a searing pain and his leg seemed to turn to Jell-O. But he'd been training for three years to run these 44 seconds. Three teammates and an entire country were expecting him to get to the finish line. His mind refused to take in any other information.

"I can't believe that I ran as fast as I did on a broken bone," he said later. "It's still unreal to me."

Your body doesn't generally want you running on a broken bone—that's the whole point of feeling pain. But after the body sends the signal of injury, it's up to the brain how to interpret it. In runner Mitchell's case, the brain's answer to the signal was *I haven't got time for the pain!*

The brain doesn't usually choose to ignore the body's signal, but it happens more often than you might think. A hockey star for the Boston Bruins named Jake DeBrusk scored two goals against the Pittsburgh Penguins during the 2023 Winter Classic, one of the National Hockey League's premier events, and both of the goals (including the game winner) occurred after a puck slammed into him and seriously injured his leg. He knew something was wrong but "didn't say anything," he admitted later. The injury eventually had him sidelined for six weeks.

Another Bruins player stayed on the ice for almost a minute after breaking his right fibula in a 2013 Stanley Cup playoff game—also against the Pittsburgh Penguins. The injury occurred in the midst of Pittsburgh power play, and the injured player, Gregory Campbell, waited until the penalty period was over to get off the ice. I suppose one lesson from all this is that Bruins-Penguins games can be dangerous. More important, though, is the lesson that while we think about pain as coming from the body, the *response* to an injury or other pain is fully in the brain.

Playing through pain is rarely a good idea, but I find these cases fascinating because they present incontrovertible proof that our body's pain signal is only a small part of our experience of pain. In certain situations of high tension or danger, the pain alert from a hurt body can get ignored by the brain. The phenomenon, known as stress or threat analgesia, evolved to protect you when something particularly dangerous might be going on. If you're facing an attacker or some other immediate threat, "you don't want to be hobbled by aches and pains. You want to dampen those things down and act with full strength in the moment," Tor Wager told me. Attacked on a dark street corner by someone wielding a knife, you might not even realize your arm is slashed until you've run away and are someplace safe again. At that point, the dripping blood and rush of fear may make the cut hurt even more.

Since I grew up outside Boston with a dad who loved hockey, my first instinct in hearing about the heroic Bruins players was that these guys are just tougher than the rest of us. Pain or a broken leg? Bah, humbug! But as I investigated further, I realized that the stories revealed something subtler going on. The athletes' bravery might have had less to do with a conscious decision than with the interesting ways that body and brain intermingle their signals to decide how to proceed.

Most of us won't be involved in knife fights or the Olympics or Bruins games, but it's interesting to think how we can use these episodes to change our own moments of pain.

We know that worrying about pain makes it worse. When you ruminate, going over the same problems in your mind again and again, the difficulties start to loom even larger. This can happen with any obstacle, from getting divorced to losing your job. You begin imagining the terrible possible outcomes and focusing on all the distress and anguish ahead. The more you perceive threats and adverse outcomes, the more negative your mindset becomes. Your body responds by revving up and tensing as it would for any threat. When the core problem is already a physical pain, the fear just makes the pain more intense and intractable.

If you're already a worrier, I apologize for adding to your anxieties. But this is one case where it helps to believe what you read. Once you are able to recognize your own tension and the effect it has on your pain, you can start to change the pattern. Many pain clinics now offer behavioral therapy to help people think differently about pain and put it in a more reasonable perspective. Vast amounts of research show that it works. As we saw in Tor Wager's study, a few sessions of cognitive behavioral therapy can help you break the cycle of rumination and fear that can create chronic pain. Many patients are now encouraged to view their pain as a neutral sensation coming from their brain—not a danger signal from their body. In one study, an impressive 73 percent of the patients taught to see their pain as non-threatening ended up pain-free, with improvement that lasted for at least a year.

If you want to resolve the pain, you need to trick your brain into doing the opposite of worry and rumination—and there are some techniques you can try at home. From the mounds of research I've looked at on pain, I've come up with three ways to help you think about pain in new ways—and maybe make yourself feel better. You can try to distract your brain, you can change expectations, and you can work on your perspective. Here are a few ways to proceed.

LOOK OVER THERE!

One of the simplest ways to change your pain experience is through distraction. The human brain is regularly extolled as a great marvel, and I don't doubt it. But let's be honest—it has some limitations. One of them is that it can't do too many things at once. When you think you're multitasking, you're probably just switching from one task to the other and doing less well on both of them. One study found that only about 2 percent of people are actually able to dual-task without suffering significant changes in performance. The same results have been repeated often. We all like to think that we're among the 2 percent of "supertaskers"—but come on. Odds say you're one of the 98 percent who do worse when attention gets divided.

When it comes to pain, your brain also isn't very good at dividing its attention. One of the great mysteries of pain has often been why it gets worse at night. Lie down to go to sleep and suddenly those back aches get worse and the neck twinges hurt more and the stomach pangs you'd felt earlier seem to magnify. You can find lots of body-related explanations that involve the right mattress or the wrong pillow or the perfect position for lying down. It's far more likely that your body is continuing to send the same signals it has all day—but now your mind has no other distractions. Lying in your quiet dark bedroom, you're not getting any new sensory inputs or distractions, so your mind focuses strictly inward. In trying to figure out what those bodily sensations mean, your mind amplifies them. The neural circuits that involve anxiety and worry start to overlap with the pain circuits. You can get as many new pillows as you want (and trust me, I've done that often), but it's not going to help the real problem.

Distraction can be a great analgesic (or pain reliever). A lot of research shows that if you turn on some music, you're likely to turn down the amount of pain you feel. Music changes how people perceive pain so dramatically that playing music might make someone point to a 3 on the pain chart rather than their usual 5. The effect is clear, but the reason is still up for grabs. Healthy people put in MRI machines show changes in neural activity while listening to music, including in brain areas known to be involved in pain transmission. I've also read studies linking music to changes in levels of catecholamine, a chemical made by nerve cells that's involved in the stress response. Music may lower your stress and anxiety, which is why your dentist probably plays endless loops of mood music while you're in the chair. (Adele and Ed Sheeran seem to be favorites.) On the most straightforward level, music may help relieve pain by diverting your attention. When your brain has a competing stimulus, it can't focus quite as much on the drilling taking place on your left molar. Or as reggae singer Bob Marley explained it, a good thing about music is that when you hear it, you feel no pain.

Other kinds of sensory input can also distract your brain from pain. The use of essential oils to treat pain seems to date back to medieval Persians—and they were probably on to something. Now we call it aromatherapy, and I've seen studies showing that lavender oil or orange oils can lessen pain during childbirth and others suggesting that aromatherapy with ginger oil is effective for chronic back pain. You can also find advocates of rose oil, bergamot, wintergreen, peppermint, clary sage, clove, lemongrass, and many others. One study that evaluated more than three hundred research papers concluded that aromatherapy can be effective in making patients happier but "the reasons for these results are unclear." Maybe the rituals around aromatherapy are soothing enough to reduce stress and anxiety. Or maybe when you smother your senses in the rich smell of lavender, your mind has a new distraction that diverts some attention from the pain circuit.

Simply directing your attention elsewhere—or having someone help you do that—can redirect the pain circuits. At a children's soccer game one day, I watched a little boy named Rowan get knocked over and begin to cry. As the coaches

Pain, Pain, Go Away

rushed over asking "Are you okay?" and "Do you need ice?" his sobs just got louder. His dad walked over, and with a concerned look he held out two fingers. He asked Rowan to squeeze the fingers as tightly as he could. Rowan wrapped his small fist around the dad's fingers and complied. "Tighter," the dad said. As Rowan focused hard on the squeezing, his crying subsided. Then the dad had him do it with the other hand, and he gave a few other directions that Rowan followed. Finally the dad gave him a hug. "Fortunately, I think you're okay," he said. The now-calm five-year-old ran back into the game.

Impressed by the exam, I went over and asked the dad (who I assumed was a physician) what he had been testing for. It turned out that it hadn't been a medically focused intervention at all. "I wanted to see if he was hurt or just scared," he said. "The squeezing game got him distracted enough that he forgot about the injury—so I had the answer."

"Very smart. You must be a great doctor," I said.

"Actually I'm a cellist," he said with a smile.

I don't know if the dad's musical skills provided his parenting insight, but his analysis was exactly right. Most pain ends on its own, and a bang on the shoulder (like little Rowan got) or a small cut or pinprick resolves in a few minutes. Even a muscle pull or a torn ligament will usually heal in a matter of days or weeks. Longer-lasting pain can often mean the neural circuits in the brain have gotten in a tangle. The well-meaning coaches asking about ice and pain could make the brain focus on the injury, creating anxiety and fear that just exacerbate the problem. The father understood about distraction. Playing his son Haydn's Cello Concerto in C Major might have had the same calming effect. It just wouldn't have been quite as easy on the soccer field.

DON'T MAKE A MOUNTAIN OUT OF A MOSQUITO BITE

Another secret for handling pain is to change your expectations. How we think about pain changes how we experience it. If you know something is going to be short-term and self-limiting, you can usually handle it much better than if you fear that it might go on and on. When you don't know exactly what's happening, your fear sets up a cycle of worry and tension that makes the situation worse. Psychologists refer to it as pain "catastrophizing"—a scary word for a fairly common cycle.

It begins when something hurts and you wonder how serious it is and if it's going to get worse. In the next stage, the concern makes you ruminate even more—thinking about how much it hurts and magnifying the pain. You start to focus on it intently. That can lead to the final stage, where all the worry makes you start to feel helpless that there's nothing you can do and the problem will never end.

When you ruminate, going over the same problems in your mind again and again, the difficulties start to loom even larger. This can happen with any obstacle, from getting divorced to losing your job. You begin imagining the terrible possible outcomes and focusing on all the distress and anguish ahead.

The more you perceive threats and adverse outcomes, the more negative your mindset becomes. Your body responds by revving up and tensing as it would for any threat.

When the core problem is already a physical pain, the cycle of rumination-magnification-helplessness *does* make the pain worse—so you are actually making your fears come true. Many studies show that people who catastrophize experience greater pain intensity—because they are unwittingly feeding their own pain experience.

We've all fallen into the catastrophizing trap. It might start when you have an itchy red bump on your arm that seems to appear from nowhere and doesn't go away. You search online medical sites for information, holding your arm up to your iPad to compare your bump to the pictures of various maladies. Could it be a basal cell carcinoma? At least that's the curable kind of skin cancer. You make an appointment with your dermatologist but she's not available for two weeks, so in the meantime, you look at more online pictures and think about the spot while you're lying in bed. Eventually you decide that you've been fooling yourself and the strange spot is a melanoma—the more dangerous kind of cancer. Yes, that's what it is. Melanoma. You might not survive. You think of your sweet children lying in the next room who will have to grow up without a parent. Tears fill your eyes and the dreaded spot seems to be throbbing, keeping you awake. Now you start to have a stabbing pain in your stomach too. Maybe the cancer has spread.

This situation happened to a friend of mine last summer.

By the time she got to the dermatologist, she had no doubt what she would hear. But after the dermatologist examined her, he sat back thoughtfully and offered his diagnosis.

"It appears to be an infected mosquito bite," the doctor said.

My friend is able to laugh now when she tells the story—and she always adds that the stomach pain disappeared in the doctor's office too. In general, she's not a catastrophizer, so she was able to see what had happened. She was essentially causing her own pain.

DREAM OF A MAGIC PAIN PILL

When researchers study people who are about to undergo surgery, they can reliably predict which ones will have the worst post-operative pain. Want to guess what they look at? No, it's not the severity of the problem, the patient's general physical health, or even the skill of the surgeon. Instead, the people who end up with the most pain after surgery had the highest levels of fear and magnification *ahead* of time. In studies of people about to have musculoskeletal surgery—such as on a knee, shoulder, or hip—the pain worriers are also the most likely to develop chronic pain down the line. Fearing that you'll have a lot of pain and a bad outcome just makes it more likely that you will.

If I've now made you worried about worrying too much, please know that you can turn that around. Lowering your fear and believing that you can handle whatever occurs will change your experience of pain in a positive direction. I learned that

Pain, Pain, Go Away

my sophomore year in college when I contracted a viral infection that brought a very high fever, painfully enlarged glands, and a throat so inflamed I couldn't swallow my saliva. Being generally stoic, I remained curled up in my dormitory room until my boyfriend at the time got alarmed enough that he ignored my objections, scooped me up, and whisked me off to get care.

Finally lying in the university hospital, alone and feeling very sorry for myself, I agonized over how I could possibly handle the worst pain I'd ever had. I started spinning imaginary stories. (This is what happens when you major in creative writing.) I fantasized having a magic pill that would immediately end all pain. Before I reached for it, I added a plot twist. The pill could never be replaced or used again. I would get exactly one magic pill in my whole lifetime.

Would I use it up now?

I desperately wanted to get back to my own dorm room (and the nice boyfriend). But I was only nineteen! Surely at some point in the future, I would be struck with a pain or illness that would make my current experience seem insignificant. Maybe childbirth would be awful or I'd be in an accident or…I couldn't even come up with all the terrible possibilities that might befall me in a lifetime, but I solemnly decided I would not take the magic pill now. It had become very real to me, and I imagined how grateful I would be down the road that I had saved my magic pill.

I must have been half-delirious to come up with this

fantasy scenario, but it was also oddly comforting. However terrible I felt, I accepted that I could get through it. Things could be worse. The virus would run its course (it did), and the pain and inflammation would subside.

The magic pill has stayed in my arsenal ever since, and it has always served the same purpose. *I won't take it now! I can handle this!* I remember myself lying there at nineteen and realizing that I could survive the pain—and I know that my current situation (whatever it may be) will also get better. My expectation that this really isn't so bad and could be much worse keeps the bubble of pain from blowing up too high. It gives me back some sense of control, and that in itself eases the current pain.

What Doesn't Work

I should note that while my magic pain pill was imaginary, the search for a real one has brought horrible outcomes. The opioid epidemic that has devastated so many parts of our country began, at least in part, from a genuine effort to get pain under better control. Among the many other frauds and tragedies that resulted, one of the most unexpected is that while opioids have some short-term effects on pain management, in most cases, they don't perform much better in the long run than other drug treatments. In one workshop run by the National Institutes of Health, researchers looked at the many treatment options for pain and considered both the benefits and dangers of opioids. The executive summary concluded that "there appear to be few data to support the long-term use of opioids for chronic pain management."

Drug therapies seem to work best for the "nociceptive" pain, like that resulting from injury, arthritis, or cancer. But opioids and other pharmaceuticals seem to have very little effect on the conditions that arise from central nervous system pain syndromes, including the chronic pain of migraine headaches, aching backs, fibromyalgia, and irritable bowel syndrome. Those respond best to behavioral therapies and physical therapy and alternative treatments like acupuncture. In other words, it's often the case that an imaginary magic pain pill actually works better than a real one.

Opening and Closing the Gates of Pain

Even though the research about mind-body connection in pain is pretty overwhelming, some people find it either disconcerting or unbelievable. Whenever articles appear online about the value of behavioral therapy for pain or the link between tension and back aches, the comments section goes ballistic. I understand. Seeing your pain problem as structural makes it very straightforward—fix your back and the pain goes away! Unfortunately, it doesn't work that way. Pain essentially rewires the brain.

The body-brain pathways in pain are enormously complex. Consider, for example, a phenomenon known as the gate-control theory. Two researchers who were then at MIT first proposed the idea in the mid-1960s. They explained that the brain can handle only so many sensory inputs at once. Remember what we said earlier about how multi-tasking doesn't work? This is the sensory body-brain equivalent. When the nervous system is sending the brain signals along one pathway, it shuts out the pain signals trying to come through the same gates.

One day while I was working on this chapter on pain, the thumb on my left hand began throbbing. I expected that I had strained it holding my iPhone and texting for too long. Doctors call this a "repetitive stress injury," which generally means that tendons or ligaments get sore from doing something for too long. In the old days, it would be called writer's cramp. Without really thinking about it, I started massaging the painful thumb, and as long as I did it, the pain seemed to go away. When I stopped, I felt the stinging again.

I realized that I had stumbled on a very simple proof of one aspect of the gate-control theory—the idea of pain signals getting waylaid by other sensory inputs. In the case of my thumb pain, the researchers would explain that the skin has large muscle fibers that respond to touch. Anything the sensory nerves in your skin touches or feels gets sent to the brain for explanation. It also has many smaller fibers that shoot signals when there is some kind of muscle damage. By massaging my thumb, the large fibers were taking up all the space—effectively closing the gate. The pain fibers couldn't get through. It may be one reason that a back massage makes you feel better and that acupuncture often works so effectively to relieve pain. Both keep the pain from getting to the brain for a time, and breaking the pain circuit even temporarily can be the start of a longer-term fix.

Here's something to think about. If the gate works one way and shuts out signals from the body, it's very possible that the system swings the other way too. Neurobiologists think that, in some cases, that could be exactly what happens. Instead of closing the pain gate, the mind opens a new one and *generates* signals to the body.

The experience of so many pain researchers shows that people who learn to revamp their own perceptions of pain can become immediately happier with their own lives. Your body gets no benefit from chronic pain—and doesn't really want you to experience it. Understanding some of the complex interweaving of nerve pathways, emotions, neurotransmitters, and brain signals that lead to pain helps you gain an important sense of control over your outlook and daily mood. It gives you a way to take one giant step toward being happier.

13

Sugar Pills Are Sweeter Than You Think

> Everything is a miracle. It is a miracle that one does not dissolve in one's bath like a lump of sugar.
>
> —PABLO PICASSO

STANDING AT THE COUNTER AT my local pharmacy the other day, I noticed a prominent display of a dozen or more very impressive-looking pill bottles. Small signs under each one explained their individual purpose: Relax, Sleep Support, Eye Health, Energy, Knee Repair, and many more. I felt myself getting very excited. A pill to give me more energy! Another for my always-hurting eyes! I felt like I had stumbled on a treasure chest.

I reached for the Eye Health bottle first, and the very informative (and scientific-sounding) label promised that it would refocus tired eyes by decreasing cellular inflammation. Well, that sounded good. The ingredients list described a naturally occurring carotenoid (like the stuff in carrots) that was derived from something with a very long Latin name that I couldn't pronounce. I looked it up and discovered that it was a microscopic alga that has also been promoted to enhance muscle endurance, strengthen the immune system, give you radiant skin, and improve memory and cognition, which seemed like high expectations for one green, slimy organism.

When the pharmacist came back, I held up the bottle for him. "Does this stuff work?" I asked.

"A lot of people think so," he said with a noncommittal shrug.

That didn't seem like much of an endorsement, so I put the bottle down and pointed to the supplement that promised Energy, instead.

"Can you tell me about this one?" I asked.

He reeled off the ingredients listed on the label and, sensing my skepticism, gave a bit of advice. "The general rule is that if you believe it will work, it does. If you're hesitant, you might want to pass."

Supplements like these don't need to be approved by any government agency to show up on store shelves, and whether they have any scientific credibility or not, they are enormously popular. More than half of American adults take multivitamins or other dietary supplements, sometimes in large quantities, spending close to $50 billion annually on them. Marketers pour some $900 million into promoting them. Many doctors explain that the only result is very expensive urine, since the body doesn't bother holding on to the extra vitamins and minerals it doesn't need.

Researchers have studied the effects of supplements for years now, and it turns out that the measurable advantages to your health go from non-existent to negative. An analysis of research involving four hundred and fifty thousand people showed that multivitamins don't reduce the risk of heart disease, cancer, or mental decline. The dream that supplements can help you live longer or age better seems mostly misplaced—in fact, it's upside down. Men who took multivitamins in one study doubled their risk of dying from prostate cancer, and other studies showed the risk of mortality from lung cancer and cardiovascular disease also increasing with some supplements. Major medical journals have published ardent editorials urging people to stop taking supplements. My very favorite has the unsubtle title "Enough Is Enough: Stop Wasting Money on Vitamin and Mineral Supplements."

After learning all this, I felt very smart not to have bought the Energy or Eye Health bottle at my pharmacy. But then I discussed the subject with an older doctor I know and admire who has a large practice in internal medicine outside San Francisco.

"I'm glad to prescribe multivitamins to my patients when it's appropriate," she said.

"How could it ever be appropriate if the evidence isn't there?" I asked.

She gave a little smile and told me that there are many different kinds of "evidence." She believed the studies showing multivitamins don't help with specific diseases were completely accurate and well-substantiated. But she had also seen over many years that when she recommended certain supplements to improve energy or general health, people felt better.

"Call it a placebo effect if you want, but as long as it works, why wouldn't I use it?" she asked.

This much-loved doctor didn't want me to use her name because she wasn't sure the practice was completely…ethical.

In her experience, patients benefited from the simple act of coming to see her, talking through their problems, and getting some sense of control. If an overtired parent complained about low-energy, the doctor would talk to them about getting more sleep and suggest they might be experiencing a low-grade depression that could be treated with self-care. She might explain that while there wasn't much scientific proof from a physiological perspective that a B-12 supplement would improve energy and mood, she had found it helpful for many of her patients. Presenting it that way, which was both honest but positive, seemed to give additional potency to the supplement. Many patients would come back and tell her how much more energy they had since getting the B-12 or the multivitamin or whatever other supplement she had recommended.

"Sometimes the vitamin pill prescribed by a doctor gives you permission to let your body do its work of feeling better," she told me. "Our bodies have great innate healing abilities, and my job is figuring out how to release them to work."

When pharmaceutical companies want to get a drug approved, they need to show that their new remedy works better than a placebo. It doesn't sound like a very high bar, since a placebo is essentially a sugar pill with no active properties. But in double-blind studies where neither patient nor researcher knows who is getting what, placebos are so enormously effective in improving some conditions that most drugs have trouble outperforming them.

Once the drug is approved, you never hear it compared to the placebo. We've all seen too many TV commercials for pain-relieving drugs where happy people skip through fields of daisies and butterflies while a voiceover intones that the drug lowers chronic pain by 50 percent. They don't mention that 45 percent of the daisy-and-butterfly effect is coming from the pain-relieving powers of your own body.

Research by Ted Kaptchuk, a professor at Harvard Medical School who has done extensive research in placebos, shows that telling people they are getting a placebo doesn't lessen its effectiveness. In one study, he offered patients with irritable bowel syndrome sugar pills and explained that they "have been shown in clinical studies to produce significant improvement in IBS symptoms through mind-body self-healing processes." The patients who took the pills had fewer symptoms and less pain than patients given standard care. What Kaptchuk calls "open label placebo" has dramatically helped patients with migraine headaches, lower back pain, and other ailments. Placebos have also proven their power in treating depression, ADHD, and Parkinson's disease.

Any pharmaceutical company would covet that list of

successes—but a lot of people get queasy at the idea of placebos. It all sounds too magical. A sugar pill to ease your depression? An ineffective vitamin pill that can give you energy? The idea makes much more sense when you think about some of the other connections we've uncovered between body and mind. Kaptchuk says a placebo is a way for your brain to tell your body what it needs to feel better. Your body may then release feel-good neurotransmitters like endorphins and dopamine to improve your mood and help you overcome pain. These natural opioids work in a similar way to those made synthetically in a laboratory and have far fewer dangers and side effects.

Instead of being skeptical about placebos, we should celebrate them as one more example of the body's genius. You have internal resources to make yourself feel better. Your body is making itself well rather than needing an outside source—which should be reason to cheer. Our bodies know how to make us feel better and happier, and we just need to urge them to do their work.

In the fifth century, Hippocrates described laying his hands on a sick patient and being able to draw away the aches and pains. Offering an update on that view, many general practitioners believe the ritual of going to a doctor and being examined while getting hands-on care and attention is an important part of the placebo effect. The contorted systems of American health care may be undermining that positive advantage. People used to visit a physician for an annual exam, feeling reassured after a doctor listened to their heart, felt their glands, and palpitated their abdomens. Now Medicare has stopped paying for an annual physical exam, and other insurers have followed. In its place, Medicare allows a yearly wellness evaluation that includes a discussion of how to stay healthy but no hands-on exam. If you want any sense that your body is actually involved in the process of keeping you well, you have to pay extra.

"The patient doesn't get undressed, and you don't touch them or create any bond," says the San Francisco doctor who believes in the power of personal connection for better health. "We are taking away one of the key ways that the doctor-patient connection helps people get themselves well."

Why Sugar Pills Work

We've already seen that pain develops within a context of mind-body interactions. Your fears and expectations about pain and the perspective you have on how bad or long-lasting it will be all contribute to what you feel. It works the other way too. What you think will make you better often does. Neuroscientist Tor Wager (who did some of the interesting research on pain discussed in the previous chapter) described to me the yo-yo-ing that has occurred in our perceptions of placebos, swinging from "It's all just people fooling themselves" to "Placebos are so powerful they can affect everything." His mission has been to see where in that range placebos actually fit—and what they can and cannot affect.

When I spoke to him, Wager admitted with a smile that some of his early studies probably contributed to the idea of placebos as all-powerful. He and other colleagues showed that placebos can dramatically turn down pain signaling. "The placebo effect works in part by causing the brain to release its own opioids," he explained. It's quite stunning when you think about it: Given the right impetus, your body can tell your brain to release the neurotransmitters that will make you feel better. Our bodies and brains can work together to make our lives smoother without any outside chemical interference.

Once it became clear that placebos work to reduce pain, Wager began to wonder *where* in the brain and body these effects take place. He's now done large-scale studies involving a lot of fMRI imaging that show most of the effects are in the areas of the brain where the pain is being processed and interpreted. Much less affected are the sensory pathways themselves.

As an example, imagine you're given a cup of coffee to hold that's painfully hot. Your response is *Ouch, that hurts!* Then you get another cup of coffee the same temperature—but before holding it, you're given a cream to rub on your hands. You're told that the cream will reduce your sensitivity to the heat and make you feel better. If you're like most people, you'll find that holding the second cup of coffee isn't nearly as painful. That's true even though the cream doesn't really have an effect on your skin or temperature sensitivity at all.

What is making you feel better? Wager found that the sensory pathways from the skin to the spinal cord and on to the brain respond the same with or without the placebo. In other words, the actual *sensation* you experience doesn't change. But there's a change in what Wager describes as the "core central circuitry of the brain that is involved in decision-making and emotion and learning. The placebos are affecting those areas." Your brain gets the same sensation from your hand but interprets it differently.

Not being a scientist like Wager, I can make a leap to say that in a funny way, the body is harder to fool than the brain. Your body keeps doing its job and sending the signals of what you're experiencing. It doesn't get involved in value judgments. When you're told that the hand cream will make the pain less intense, the body basically shrugs off the information and continues to send the same sensory pain signal. *Can't fool me!* It doesn't do anything differently. But your brain is highly suggestible. When it receives the same sensory information but in the new context, it now registers the sensation in a completely revised manner.

Wager says that the particular motivational centers that get influenced by placebos make them notably effective in lessening depression. This is a big deal—and if we truly want to make ourselves feel better and happier, we should realize that it's a *really* big deal. More than twenty million Americans swallow antidepressants to the tune of about $11 billion a year—and as you can imagine, the big pharmaceutical players don't want to see this game disrupted. But it probably should be. Antidepressant drugs work—just ask any doctor who prescribes them or patient who takes them. Symptoms improve and patients feel better. What doesn't get discussed, though, is that some of the value of the drug—often a significant amount—comes from something *other* than the chemicals it contains. Reliable studies show that about 82 percent of the benefit of antidepressants comes from the placebo effect.

You can find passionate debates about the value of antidepressants in medical journals but not much argument about the 82 percent figure. Supporters of the drugs argue only that the extra 18 percent benefit from antidepressant versus placebo is enough to warrant taking them despite the expense and side effects. In many cases, particularly in severe depression, this small positive effect might be life-changing. How much difference that 18 percent will make in mild or moderate depression, though, is unclear and certainly a discussion worth having between patient and doctor. How often does it occur? I checked in with five people I know who have used antidepressants, and not one of them had been told about the placebo effect before getting their prescription from a psychiatrist (in four cases) or internist (in one). This is an admittedly random sample, but I expect most psychiatrists who believe in the power of their medicines don't spend a lot of time talking to patients about the large percentage of the magic that's done by your own body.

The FDA is circumspect about how much better than a placebo a drug must be to get approved, but the number is sometimes stunningly small. Two clinical trials have to show a "statistically significant" improvement over a placebo. But what works mathematically doesn't necessarily have any clinical meaning. It's like the old joke about the two lovers who are across the room from each other. One of them can move halfway closer, then halfway, then halfway again. A mathematician would correctly say that Lover #1 will never get to Lover #2. The rest of us know that for all practical purposes, they'll get there.

The popular drug Cymbalta (you've probably seen TV ads for it) was shown to be just 10 percent better than a placebo when it was approved to treat depression, anxiety, and pain. Another popular antidepressant, Zoloft, got approved with trials showing it beat the placebos by 15 percent. Both are in a class of drugs called SSRIs that supposedly work by manipulating the balance of serotonin in the body and so increasing what's available in the brain. Interestingly, another drug called tianeptine that is approved as an antidepressant in France works by *decreasing* the amount of serotonin in the brain.

In trying to explain this seeming contradiction, one psychiatrist from Harvard Medical School noted that "it simply does not matter what is in the medication—it might increase serotonin, decrease it, or have no effect on serotonin at all. The effect on depression is the same. What do you call pills, the effects of which are independent of their chemical composition? I call them placebos."

Many psychiatrists would disagree with this assessment and point out that however small the added advantage of the drug over the placebo, it can have an important effect for some patients. What's helpful to understand is that each person's body chemistry has its own particular rhythms. Medicine turns out to be much more art than science, and when drugs work, they can change lives for the good. When we can call on mind-body interactions to get happier, we should make sure to realize how powerful those can be too.

To Cut or Not to Cut

We are not very good at knowing what is making our bodies feel better, and so we attribute extra power to drugs, surgeries, and fancy equipment. In one astounding study, patients with knee pain were divided into three groups. Two of the groups got the two different arthroscopic surgeries that are commonly used for knee pain caused by arthritis, tendinitis, ligament tears, and other non-emergency events. The third group got an incision in their knee—but no actual surgery. The patients were followed for two years. They regularly reported on their own pain and function, and the researchers followed up with objective tests for walking and stair-climbing. The results left no room for doubt. In a paper published in the *New England Journal of Medicine*, the researchers concluded that the outcomes of the arthroscopic surgery were no better than the placebo treatment. Should I repeat that?

Let me say that again because it is an absolutely stunning finding. Getting actual arthroscopic surgery where the doctor inserts tiny knives and tools inside your knee to repair damage and remove broken-down bits of cartilage and tissue has exactly the same outcome as making the incisions but not doing the surgery at all.

These findings have been around for a couple of decades, but if you consult a doctor for knee pain, there's a good chance they will still suggest arthroscopic surgery. It works, so why not? People feel better some weeks after the surgery and have no way of knowing that they likely would have improved *without* the surgery too. They're happy! They bring the doctor a bottle of wine in thanks.

We are all very good at what psychologists call confirmation bias, which means that when you have a set view on a topic, you only believe the evidence that supports your opinion. We see it all the time in political discussions and daily issues. If you just bought a new car, you tend to notice the articles that mention it as being particularly safe, and if someone shows you a report that your car is poorly made, you'll just ignore it, sure that it's wrong. People who believe the earth is flat buy flat maps, and when someone shows them a globe, they say it's not real. Similarly, if you've paid $5,000 or more for a surgical procedure, you don't want to hear that the actual benefits are mostly from your own body. You tell your friends how well it worked, and when the friends suffer a similar pain, they want to get the surgery too.

Cycles like these continue, at least in part, because we trust modern medicine more than we trust our own bodies. If something hurts, we demand that doctors *do* something, and since the devoted doctors who care about making you better spent many years in medical school learning about surgery, drugs, and other interventions, they too want to *do* something. Add in the insurance companies that will pay for arthroscopic surgery but certainly not the (less risky but very effective) placebo kind, and the deal is done. We have created a circle—you might even call it a vicious circle—of doctors-patients-payers who choose to ignore the impressive research on placebos and instead advocate for outside interventions of pills or surgery that may be complicated, expensive, and unnecessary.

Understanding the power of our own bodies could lead to formidable change in what we expect from medicine and the medical community. Internists and family practitioners know that a patient who comes in with symptoms of a cold or flu probably has a virus that will go away on its own in a few days. But patients want to be cured immediately, and they regularly request a prescription for antibiotics. *There must be something you can give me!* Some doctors will take the time to explain that antibiotics are for bacterial illnesses and don't have any effect on viruses—so, no, they won't prescribe one. Other doctors realize that writing the prescription is less time-consuming than giving a science lesson in viral illnesses. It's also what the patient wants. The patient gets the antibiotic and—*they get better!* They credit the doctor with the cure, even though their improvement had nothing at all to do with the pills they swallowed (which might have given them an upset stomach).

You see why I call this a vicious circle. Next time the patient has the same illness, they want another antibiotic. If the doctor resists, they want another doctor.

I'm not suggesting that medicine never works or that we should all just munch apples and trust in the power of our own bodies. Far from it. Modern medicine can be miraculous. Penicillin and other antibiotics have saved hundreds of millions of lives and probably changed human history, as have vaccines for smallpox, polio, measles, and many other diseases. The rounds of vaccines now routinely given to babies and toddlers have made childhood death rare rather than tragically common. Kidney dialysis, insulin, and transplants save lives every day when people's own bodies malfunction. Treatments for HIV and some kinds of cancer have been life-giving and life-saving, and we should be in awe of the scientists who have devoted themselves to understanding the power of the right interventions. Some 30 percent of women died in childbirth a few hundred years ago, and advances in medical care mean that number is now about 1.7 percent in the United States. (The U.S. has the highest maternal mortality rate in the developed world, so that number should be even lower.) Childhood leukemia used to be a death sentence, and now it can be treated so successfully that most of those stricken go on to perfectly healthy lives. Our body's systems misfire, misbehave, and go haywire quite regularly and we need the dedicated doctors who try everything they can to get them straightened out again.

Placebos will not shrink a tumor or cure malaria. They work on symptoms that are modulated by the brain—but that includes all kinds of pain and far more illnesses than we may otherwise realize.

I'll Have What He's Having

The power of placebos extends far more dramatically than you might realize. Impressive studies show placebos' potency in treating migraine headaches, irritable bowel syndrome, Parkinson's disease, and all kinds of chronic pain. The medications that get prescribed for these problems work better than the placebo; otherwise they wouldn't get FDA approval. But I am awed to discover that about half the effect of some migraine medications comes from your own body's revving up its healing power rather than the drug.

I had migraine headaches for many years (fortunately, they're gone now), and I felt very relieved to know that I had a pill bottle in my bathroom if things got bad. Taking the pills invariably helped. Would a sugar pill have done the same thing—or nearly so? I have no idea, but I suspect it would. On the other hand, I probably wouldn't have had the courage to give up the medicine that I trusted. People dealing with depression have the same experience. Depression is often described as a disease of hopelessness, and having a medication with a medical stamp of approval can succeed by resuscitating hope. In our society, bottling that hope without the veneer of scientific acceptability proves a greater challenge.

Similarly, millions of children in the United States are

deemed to have ADHD each year. Most experts agree that it is wildly over-diagnosed and over-treated, but a large percentage of the children—some as young as four or five—are put on medications meant to control their behavior. Most of the little ones being treated say they can't tell the difference when they're on or off the medicine except for the unpleasant side effects, including loss of appetite and gastrointestinal upsets. Parents and teachers and caretakers, though, usually insist that they see a big difference. Who's right? Studies have shown that once children have what doctors determine is the right level of medication, cutting the amount in half and substituting the rest with a placebo causes absolutely no change. Even more stunning, it could be that the real effect of the ADHD meds isn't on the child but on the people evaluating them. When a child is taking medication, parents and teachers assume they are better behaved and tend to be more tolerant, patient, and encouraging. The positive attention helps the child flourish—and the drugs get the credit.

If you expect something to work, it often does. One eye doctor who does cataract surgery told me how careful she is in telling patients about the kind of lenses that can be implanted during the surgery. The standard lens is covered by insurance. Several newer and fancier lenses can cost the patients several thousand dollars out-of-pocket. Those lenses can have moderately better results in vision, but they can also have additional problems.

"Some patients will be happiest if they don't have to pay anything extra," she told me. "Other patients are convinced that the more they pay, the better the outcome." She tries to match the lens to the patient's expectations of what will be best. While her surgical skills are impressive, her high patient satisfaction numbers may come from her skills in making patients feel like they've made the perfect choice. How much you have (or haven't) paid and attitude of the doctor all contribute to how happy you are after the surgery.

If you're currently on any kind of medication, I don't recommend that you stop it. If it's a long-term medicine for pain or some of the other illnesses we've mentioned here, you might have a conversation with your doctor about how much might be attributed to the placebo effect. Maybe it would be worth an experiment in cutting down the amount and substituting rituals of self-care like exercise or meditation. Maybe some version of cognitive behavioral therapy will help you as much as a pill. But it's also worth remembering that many drugs are life-saving and life-changing. Our body chemistry and physiology can determine our emotional state and change our behavior. For some people, the physiological changes that a drug creates can forever improve their mental health.

We all imagine medical science as being straightforward and clear-cut. If you have a pain, you take a pill or have a medical procedure, and you get better. But we now understand that our mind and body circuits are so deeply intertwined that their effects can be hard to separate. Your body wants you to stay healthy and strong—and sometimes getting there can be sweeter than you think.

PART FIVE

Why Pleasure and Creativity Come from Within

Everything you do with your body, from doodling to dancing, can change how your mind perceives the world. Most people are more creative after taking a walk, and even robots perform better when they can have their own sensory experience. These chapters will help you explore the sources of creativity in a new way and give you a new view of your own possibilities.

14

The Neuroscience of Invention

> I want to put a ding in the universe.
>
> —STEVE JOBS

HOW MANY ORIGINAL USES CAN you come up with for a key?

Psychologists enjoy asking questions like that as a way of measuring something notoriously difficult to assess—creativity.

When Marily Oppezzo at Stanford posed the key question to a group of volunteers, she used a standard definition of creative, meaning that the answers had to be both original and practical. Opening a door with the key wouldn't count because it's too obvious, and using the key to tie your shoes doesn't tally because it's not logical or practical. But if you suggest using the key to carve the name of your murderer into the ground as you're dying, you get full points.

I've always been intrigued by questions about creativity,

so I sat back and tried to come up with my own answers. I thought of a few possibilities for the key: I could dig a hole for planting seeds, clean the mud from my sneaker soles, or hang several together as a windchime. But I struggled for more. (Scratching an enemy's car would not be original.) Then I discovered that when Oppezzo did the experiment, she had some of the people sit down for the test and others come up with their ideas as they walked on a treadmill. The group on the treadmill did almost *twice* as well as the others. Across several experiments that tested different versions of creativity, people did 81 percent to 100 percent better when walking versus sitting. The results were dramatic in how powerfully movement opened up the free flow of ideas. Oppezzo even titled her research findings "Give Your Ideas Some Legs."

It's pretty extraordinary, when you think about it—simply moving your body stimulates your mind to be more creative. Oppezzo and others have done the research proving it, but great thinkers have known this secret for a long time. American naturalist and philosopher Henry David Thoreau once explained that "the moment my legs begin to move my thoughts begin to flow." He suggested that the process felt "as if I had given vent to the stream at the lower end and consequently new fountains flowed into it at the upper." It's a lovely image—the ideas starting in the body and then flowing up to the mind.

Psychologists like to give names to everything, and they refer to the creative musing that leads to wide-ranging thoughts (like alternate uses for a key) as "divergent thinking."

It essentially means looking for multiple solutions to a problem and being able to think in unexpected ways. In addition to the alternate-uses quiz, similar tests of divergent thinking ask people to come up with unusual analogies or a list of eight nouns that are completely unrelated. (You'll be surprised at how hard it is not to repeat a category like animals or furniture or household appliances.)

The ideal bodily movement for creative inspiration seems to involve a comfortable pace that allows your body to be stimulated and your mind to relax. When your body is moving in a way that energizes but doesn't take all your attention, your mind is free to wander. Oppezzo notes that for her, running wouldn't serve the same purpose because "if I were running, the only new idea I would have would be to stop running." It's a funny line but not necessarily true for everyone. My husband has been a casual runner for most of his life, and he still solves all problems in his head when he goes out for a jog. Whatever level of movement brings you to a point of stimulation but not distraction seems the right choice for innovative ideas. Bestselling author, podcaster, and speaker Malcolm Gladwell has come up with more innovative concepts than most people right now, and it's probably not a coincidence that he goes out for a run almost every day. "I run by myself and without music or any accompaniment," he says. Though he was competitive when young, he's no longer trying to see how fast he can go. "I'm letting my mind wander and thinking about various random things."

When I give speeches to large audiences, I always use a clip-on microphone so I can wander around the stage, and the words and ideas seem to flow much easier that way. When I'm constricted behind a podium, my mind is impeded too. Some of my best ideas for books and articles have come when I'm walking or swimming or otherwise active. In the days when I wrote novels, I always jumped on my bicycle when I got stuck on a plot point and raced around local roads for an hour or so. The joy of pedaling fast with the wind whistling through my hair usually freed my mind to figure out what happened next (or who the killer was, when I wrote mysteries). Creative ideas can often be like puppy dogs. Approach them too directly and they may run away from you, but if you focus very soft attention on them, they may cuddle right up to you.

Why You Should Wear Out Your Rug

The history connecting movement and creativity goes back to the ancient Greeks. In the great fresco work *The School of Athens*, painted in about 1510 by Raphael at the Pope's library in the Vatican, the important figures of philosophy and science are gathered in one place. You can find Socrates, Pythagoras, Euclid, Ptolemy, and many others talking and studying alone and together. (If you think they didn't all live at the same time, you're right—this is an artistic interpretation.) The two figures in the center under a main arch are Aristotle and Plato. Art historians have written extensively about the meaning of the

The Neuroscience of Invention

books the two philosophers are carrying and the directions in which their hands point, but what always intrigues me is that in the midst of this bevy of intellect, they are posed *walking*.

Aristotle led a school of philosophy that came to be known as the Peripatetic School. The story goes that Aristotle walked while lecturing, and his eager students and disciples followed behind. Some suggest that the name might have derived from the covered and colonnaded walkways (*peripatoi*) at the lyceum where he taught—but I'll go with the walk-and-teach theory. Raphael clearly believed it, and why not? The connection Aristotle made between walking and thinking has been proven right over and over again. If Raphael used their walking in the painting as a metaphor for the movement of ideas, he nailed it.

Aristotle and Plato, the walkers in the fresco painting, are the Big Idea guys who knew how to think in unexpected ways. My longtime friend Scott only jokes about being a philosopher ("a stitch in time is better than a bird in the bush," he told me recently), but he is definitely a divergent thinker. He became enormously successful by always coming up with unexpected approaches to problems—an ability both valuable and rare in his typically constrained field of law. When he got a corner office at one Fortune 500 company, the CEO teasingly asked him to put his law degree on the wall; "otherwise nobody would believe you passed the bar." Whenever he got praised for thinking outside the box, Scott got a quizzical look on his face and said, "Is there a box?"

For as long as I have known Scott, he has been a pacer. Visiting him in his office, I would watch him trek back and forth and back and forth across his carpet as he talked on the phone to colleagues or negotiated with clients. In the days before cell phones, he became an early adopter of speakerphones—the only way he could keep up his (literal) pace. After reading Oppezzo's study, I called Scott to ask if his constant motion had been a key to his innovative style.

"I'm pacing right now as we talk without even thinking about it," he said. "Perhaps it makes me more intellectually energetic." He noted that he was glad to be considered in the same school as Aristotle and that the carpet behind his various desks over the years always got worn to the floor. "Pacing has the by-product of being good exercise during what is normally a very sedentary occupation. It's also good for rug merchants," he joked.

A body in motion stimulates the brain to be in motion too. But it's not just any movement. In the extraordinary way that metaphors spring to life in body-mind connections, activity that is flexible and fluid encourages those same qualities in your thinking processes. Two researchers in Germany came up with the idea to have some people walk wherever they wanted around a large room while others followed a prescribed path. They wondered whether free walking would have different results than being told where to go—and sure enough, the unrestrained walkers did better on the various divergent thinking tests. They were more creative. They invented more

The Neuroscience of Invention

uses for (among other things) a brick, garbage bag, and spoon than the others. They came up with more analogies.

Being meticulous, the researchers had some people follow the route of the free walkers to see if the path itself might matter—but it didn't. Barbara Handel, from the internationally respected Julius-Maximilians-Universität of Würzburg in Bavaria, concluded that being unrestrained and able to move however you like helps in creative invention. "The important thing is the freedom to move without external constraints," she said. She and her co-researcher also tested volunteers in a couple of different sitting positions, and the worst performance of all came from those in "restricted sitting"—stuck in a chair at a fixed distance from a computer screen. In other words, the position most of us take when we try to create is the very worst way to be creative. When you are hunched over and staring at one image close at hand, your body feels closed off and defensive. It puts you in a protective and guarded state rather than an open and inventive one. Innovative ideas may still pop into your head—but you're not providing the best or most likely grounding for that to happen.

One of the most inventive artists of the twentieth century, Jackson Pollock, famously put enormous canvases on the floor of his studio and walked around them while he poured, flicked, and dripped paint from a can. In the few scratchy black-and-white videos of Pollock at work, he almost seems to be dancing—his feet moving gracefully, his arms loose—and you can practically feel the energy being transferred from body to canvas. Several of those paintings have sold for hundreds of millions of dollars, a tribute to the power that comes from linking whole-body movement to creative invention.

Art critic Harold Rosenberg coined the term "action painting" to describe the American artists like Pollock, Franz Kline, Willem de Kooning, and others who understood that the physical act of painting could translate to a dynamism on the canvas. What better way to portray the energetic movements of the modern world than by making grand gestures while painting. Looking at their works, you can almost feel artists' arms flying with brushes in hand or the paint dripping with gusto from the cans. Pollock's gracefully fluid motions might have inspired some of the free association that he translated to paint and enamel. Many decades after these artists made their indelible (and museum-worthy) mark, two academic researchers did an experiment where they had some people move their arms in a loose, curvy manner while others followed a rigid, angular pattern. They discovered that the physical experience of fluidity led to more imaginative responses and mental flexibility.

Learning about the movement-creativity connections, I started to wonder if we should all spend more time dancing to release our creative instincts. You don't need to go to a club with a fancy deejay or take salsa lessons—just turning on music in your own bedroom and dancing freely for a few minutes can give you an upbeat and creative boost. The freeform dancing will likely unleash more innovative thinking than a turn at a barre class. One study looked at three different groups of professional dancers with very different styles—ballet, modern/contemporary, and jazz. Using a range of tests of creativity, the researchers found that the modern dancers scored highest and the ballet dancers the lowest. Why the difference? They point out that modern dance encourages improvisation and free movement while ballet dancers follow more rigid and structured choreography. When your body is required to conform to specific scripts, your mind likely follows. Move freely and your mind expands.

I expect that the ballerinas who perform the popular *Swan Lake* on stage would question the findings and note that their whole lives are based on creative expression! But apart from the prima ballerinas, think about the *corps de ballet*—the driven and determined dancers who form the core of the company. As flocks of swans on stage, they need to keep their movements perfectly synchronous and their bodies in pristine position so they appear as one unit. Each toe point, head flex, and arm extension is predetermined and inviolable. As one of the dancers from the Royal Ballet explained, "If you get noticed for being an individual, you're doing something wrong." The dancers' physical movements may be gorgeous (and they are!), but since their bodies need to be contained and controlled, their mind's pathways likely reflect that approach too.

The ballet dancers may be better at what is now called convergent thinking. Unlike divergent thinking, which involves wide-ranging solutions, convergent thinking is important when you need to find a single, concrete solution to a problem. You need it if you're solving a math problem, answering a question on a multiple-choice test, or dancing in the *corps de ballet* of *Swan Lake*. Interestingly, in Oppezzo's experiments, walking did not help convergent thinking and even undermined it. The wandering thoughts and openness to innovation that physical wandering encourages don't help when you need to narrow your focus to one answer and get it right. Artists and philosophers and physicists and entrepreneurs can benefit from walking or moving freely in order to inspire innovative ideas, but be glad that your bank teller is sitting on a chair when counting out your money.

Convergent and divergent thinking exist on a continuum—you don't exclusively have one or the other—and both have their advantages. I find it thoroughly amazing that how you move your body helps determine which kind of thinking is likely to be your bailiwick. The mind-body circuit works both ways, of course. It may be that someone who feels more natural as a divergent thinker chooses to become a modern dancer rather than a ballerina or an action painter rather than a printmaker. But moving your body in a different way can change the creative way your mind works. Knowing that makes me want to jump for joy—and hope I'll get new ideas along the way too.

The Cerebral Mystique Is a Mistake

I recently went with a friend to a vast art gallery displaying several dozen pieces of very modern paintings. Standing in the brightly lit space with its soaring ceilings and surrounded by the colorful art works, we both felt slightly giddy. Our bodies took in the color, light, and openness and sensed something positive going on. We talked with great animation and rising levels of excitement.

"We have to buy one of these!" my friend said with enthusiasm. I quickly agreed, and we walked around feeling slightly elated.

When we finally left the gallery, we started to calm down—and as we finally sat down for lunch, my friend wondered what had come over her. Buying one of those expensive pieces now seemed like a crazy idea.

"I felt like I was on drugs in there," she said.

Nobody had spiked our Perrier, but I expect that our levels of dopamine had been nudged upward by the positive physical sensations we experienced. Our bodies didn't care about the high prices on the paintings—they simply responded to the color and light and space. Once our hormones reacted to the upbeat sensory experience, our minds got hijacked too. Would we even like any of the paintings if we took one home to a less exhilarating space? Who cares. When your body says yes, your mind has a hard time saying no.

What happened to us in that art gallery was a great example of the body hijacking the mind. Alan Jasanoff, a professor of biological engineering at MIT, says that our bodies get flooded with the equivalent of about 10 megabytes of information every second. (Since a megabyte is 1,000 bytes, that's a lot of stimulation.) The presence of colors and light levels in our environment—what my friend and I experienced in that art gallery—can have a huge effect "both on the emotional aspects of personality and on cognitive function," Jasanoff says.

Jasanoff is affiliated with the McGovern Institute for Brain Research at MIT and heads up a lab that is developing new methods for brain neuroimaging. All of that makes it even more fascinating that he thinks we over-glorify the brain. He describes the brain as "a grimy affair swamped with fluids, chemicals, and glue-like cells called glia." Instead of mythologizing the brain we need to recognize that it is biologically based. What Jasanoff calls our "cerebral mystique" makes us forget that the brain is just an element in our personality, our creativity, our imagination, and even our cognitive abilities. Jasanoff says that our brains "are complex relay points for innumerable inputs, rather than command centers."

The cerebral mystique rises in full glory around discussions of innovation. We hear often that great discoveries occur from some "Eureka!" moment—that inexplicable instant when you spontaneously grasp something vital and previously unknown. The Greek mathematician Archimedes is said to have had his great breakthrough as he sat in a bathtub, and he was so excited that he jumped out and ran naked down the street to tell the king. (*Eureka!* means *I have found it!*) The image of the naked guy discovering one of the great principles of physics has lasted with us for well over two thousand years. But we tend to forget that the bath isn't just a funny sideline to the story—it's the whole story! Archimedes got his insight when he sat down and water rose up. It's not a coincidence that he was surrounded by the water—the physical experience led to his understanding that the displaced water was an exact measure of his volume.

Another beloved story of discovery has Isaac Newton sitting under an apple tree in about 1665 and suddenly getting the idea for gravity—then connecting that to an insight on what keeps the moon and planets in their orbits. Sheer brilliance on display while he was just sitting there! But once again, the idea didn't pop into his mind completely unbidden. He watched an apple fall and wondered about the force that made it go straight down rather than sideways. The physical experience led to a creative breakthrough. In the spontaneous imagining of new ideas, the brain doesn't work alone. An outside stimulus often can be key.

Archimedes' bathtub *Eureka!* and Newton's apple have been told so many times over the centuries that they are engrained in our collective imagination. I wouldn't be surprised if they didn't happen exactly as told—time and memory tend to embellish legends—but the stories resonate at least in part because they make sense. We understand the idea of something happening to us that alters how we think, of a physical experience changing the creative process. It would be much harder to believe a story of Newton sitting in a windowless room and suddenly understanding the planetary orbits. Great theoretical ideas often start from a substantive grounding.

Many practical inventions also involve a physical event that prompts a surge of divergent thinking. One day in the 1940s, engineer Percy Spencer noticed a candy bar in his pocket had melted into a gooey mess. He had spent years working with magnetrons, a tube that generates microwaves and was used in radar systems during the 1940s. He guessed that the magnetrons might be responsible for his squishy chocolate—and he immediately started thinking how they could be used to heat food on purpose. He put some popcorn kernels next to one of the magnetrons and *voilà!* The kernels popped. The first microwave ovens were built shortly afterward. (It took about twenty years for them to be small and cheap enough to be used at home.)

Similarly, an inventor named George de Mestral developed Velcro after he found a bunch of burrs stuck to his dog's fur and his own clothes after a walk in the woods. Trying to figure out what made the burrs stick, he studied the tiny hooks on them under a microscope and got the idea for a fabric fastener. Next time you secure a toddler's sneakers without fussing with shoelaces, you can think of him.

Not everyone would look at melted chocolate or a grungy dog and come up with microwave ovens or Velcro. But if you want to come up with the next great invention, don't expect your brain to spew it out without your body getting involved too. Go to an art gallery, swing your arms, or take a bath. Your body likes sensations and needs to share them with the brain for the unexpected to happen.

How to Learn Words Underwater

Our bodies contribute to our creativity, and they get involved in our straightforward thinking and problem-solving too. Even topics that sound like they should be exclusively handled by our brain—like math problems or vocabulary words—are also influenced by our bodies. You'd think that you should either know the answer or not, but it doesn't quite work that way. We are all complex beings, affected every moment by millions of sensory inputs. You can teach a computer that *flabbergasted* means "overcome with surprise or astonishment," and it will know it forever. But not so the rest of us. What your body is experiencing at any given moment can affect how well you learn something and whether you remember it or not (a fact I was flabbergasted to learn).

In one experiment, researchers put people into two groups and had them learn a list of words. To make it more interesting, one of the groups studied the words on land and the other donned scuba gear and learned them underwater. (All were competent divers, but really, how do researchers come up with these things?) The participants then had the chance to be tested on both land and water—and they did better at remembering the words in the environment where they'd learned them.

The lists learned underwater were recalled significantly better when the divers went back underwater, and the lists learned on dry land got the best scores when the divers stayed dry. The results gave strong support for what they called "context-dependent memory." The place you are when you learn, hear, or discover something becomes part of how you remember it.

Other studies not requiring goggles and oxygen tanks have supported the findings. Our brains do best at encoding information when it's tied to a specific place or physical experience, so where you learn something contributes to how (and if) you remember it. If your child is about to sign up for the SATs, it might be a good idea to take the test at their own school rather than another one. Many variables influence test-taking ability, but the home-team advantage seems to work for your brain even more than it does for the local basketball team.

Given that the variables of where you take a test can affect the outcome, it seems an exaggeration to call these "standardized" tests. They are not quite as consistent in results as we may imagine. Our cognitive functions can be affected by everything from the light in the room to the comfort of the chairs. When it comes to personality tests, the results are even more inconsistent. Millions of people have taken the Myers-Briggs test, which is required by many companies, governments, and military units. After a series of questions, it confidently slots you into preset categories, announcing if you're a thinking or feeling person, an extrovert or introvert, judging or perceiving, etc. It all sounds very fixed and certain—but it is supported by zero scientific evidence. In fact, the results are so random and inaccurate that if you don't like the findings one time, just take the test again and a new you with a new personality profile will likely emerge. One study found that 50 percent of people who retested with just a five-week gap fell into different personality categories. Another study put it at 75 percent.

Why do we keep believing in these faulty tests? I think it goes back to the misbegotten idea of the "real you." Myers-Briggs pretends to uncover the one unchangeable self that defines who you are. But unless you are mummified and taking in no new physical input, that single self does not exist. Your body and brain quite literally change with the tides—as well as with the light, the sounds, and the movement around you. Being in a beautiful, natural setting increases both creativity and happiness. Most people also have an uptick in both well-being and creativity when they are in light and airy places with high ceilings and distant views. If you are stressed, go to a place where there is lots of green (including the outdoors), which is generally the most relaxing color and also inspires creativity and originality.

How many uses can you come up with for a key? Depending on how your body feels in the moment, the answer may be... more than you think. Your body can make you smarter in many more ways—as I was about to discover.

15

How Your Body Makes You Smart

The mind is not a vessel to be filled but a fire to be kindled.

—PLUTARCH

I'VE ALWAYS BEEN A DOODLER. I regularly reach for a felt-tip pen during phone calls and create elaborate geometric patterns of grids and circles. It happens during large in-person meetings too. At a conference I attended doing research for this book, I got a cup of tea before the first talk then used the paper napkin to absorb my neatly shaded doodle of lines and squares. A colleague sitting nearby teasingly asked if I planned to frame it. I'm not a painter and can't draw a horse or face in any way that you'd recognize, so I've never been sure what causes my deep drive to doodle. I would have guessed that the pattern-making was a distraction to keep myself from being bored. But now I'm happy to realize (and to let my colleague know) that it is quite the opposite. Doodling helps me focus on what is being said.

A UK psychologist did an experiment where she had people listen to mock telephone messages. Some of them doodled by shading shapes on a piece of paper while others just listened to the messages. In a surprise memory test afterward, the doodlers did 29 percent better at answering questions about the content of the messages. Doodling had helped them focus better and remember what they heard.

The psychologist suggested that the doodling might have kept people from daydreaming and so they paid more attention. But a more involved study led by art therapist Dr. Girija Kaimal at Drexel University shows something even more complex going on. She used an imaging technique (called functional near-infrared spectroscopy) to look at brain activation when people doodled, colored, or did free drawing. She found that all three of the visual arts resulted in "significant activation of the medial prefrontal cortex compared to rest conditions." Doodling brought the highest activation, including to the brain's reward centers. The people who doodled also felt better afterward about their creative ideas and their ability to solve problems.

How is it that doodling makes you happier and more focused? Here's one possibility. When you move or change your environment, your visual sense sharpens and your mind becomes more alert. It's one of those twists of body-mind interplay that make sense for our species. When very little is changing around you, your brain gets lulled into inaction. No stirring on the street or change in the environment means no threats to worry about so your mind naturally stops focusing or paying quite as much attention. If you touch something and stir up new sensations, your mind becomes sharper and brighter. A reason to pay attention! We fiddle with our fingers and rub fingers on a table in order to feel something with our bodies—and that increases brain activation.

Instead of distracting from what else is going on, doodling makes you more aware, focused, and eager. When I doodled at the meeting, the rhythmic motion of my hand moving allowed my brain to concentrate. By doodling, you're creating something that sparks your mind, and visual excitement makes your brain more alert. Doodles may also make you happy—you've created something fun to look at and you can be secretly proud. Nobody is judging your art skills (or lack of them)—it's just a doodle! The activity helps your mood improve and gives you a more positive attitude toward that otherwise boring meeting or phone call.

Our bodies weren't made to sit still and take in information. Moving, creating, drawing, and gesturing help us think better. You might say that our brain's first job is to protect the body, and the body's job is to stimulate the brain.

I, Robot

If we invented and created only with our brains, it wouldn't matter whether we sat or walked or doodled or did somersaults when thinking—but a brain never operates on its own. We have gotten used to the disembodied voices of Siri and Alexa and Google, but computer scientists mostly agree that they're not the best model for intelligent AI. Computer scientist Rolf Pfeifer has said that "intelligence always requires a body."

I went to my friendly ChatGPT, the artificial intelligence created by Microsoft and OpenAI that has gotten so much attention, to ask what it thinks about needing a body to be smart. ChatGPT is a large language model, which means that it has been fed vast amounts of text and data and essentially selects what is most likely to be the next word in a sentence. But when you ask it a question, it responds so quickly and with such personality that it feels a lot like you're talking with a sentient being. It can feel like the biggest brain you have ever encountered—able to answer anything you ask in seconds.

"Do you think you would have different responses if you had a body?" I asked.

"Having a physical presence could potentially enhance my abilities," the AI immediately replied. "It might influence the way I provide information or the examples I use. For instance, if I were able to experience sensations like touch or taste, I might be able to offer more detailed insights in those areas."

Even though I *know* ChatGPT doesn't have human emotions, I had to smile. I could almost hear its sigh of disappointment at being disembodied. Never to be able to touch or taste! Somehow this AI knew (as much as AI can be said to "know") that it could be better if only it could have sensory learning of its own.

Most experts in the AI community would agree. Giulio Sandini, Director of Research at the Italian Institute of Technology, says that "If you want to develop something like human intelligence in a machine, the machine has to be able to acquire its own experiences." All the things we do with our bodies, the smells and sounds and tastes we gather, become part of our memories. We learn by using our bodies and then use that information to understand what will make us happy.

"I don't think that one can build a representation of an apple with taste, size, shape, and smell solely through theoretical measures," Sandini says. "An apple is all these things that you have experienced yourself. You cannot preprogram an intelligent system."

The newest robots are being developed with sensors so that they can act reflexively rather than waiting for a programming response. Roboticist Rolf Pfeifer says to imagine if instead of fingertips, you had thimbles on all your fingers. Now imagine trying to pick up a glass. Without the exquisite feedback that our fingertips provide, the process would be much harder. In fact, some of the decision on how to hold the glass is handled by your hand without input from the brain—the sensors in your fingertips choose the right position to take and the amount of pressure to apply. The thimbles couldn't do that.

The same problem arises when a computer tries to tell a lever or robotic hand what to do. A computer brain can give instructions to a robot on how to hold a glass, but unless the robot has sensors that provide physical information too, the process is going to be rudimentary.

"The brain is not the sole and central seat of intelligence," writes Pfeifer. "Intelligence is instead distributed throughout the organism."

A longtime adage of design holds that form follows function. We now understand that form also *creates* function—as well as intelligence and much of the pleasure and experiences we take from the world. We sanctimoniously toss around phrases like "Don't judge a book by its cover"—but how does that make any sense? The physical substance of anything—a book, a robot, a flower—makes a huge difference. Whether you are holding this book right now in a hardcover edition or reading it on a phone, Kindle, or iPad can change your sense of its content. How the book feels in your hand and how the type hits your eyes all affect the impressions you have of it. When an auction buyer at Christie's spent some $10 million for a first edition of Shakespeare's plays a couple of years ago, they weren't looking to be inspired just by the Bard's words. The same content appears in a ten-buck paperback or for free online. First editions of Jane Austen's *Pride and Prejudice* and JRR Tolkien's *The Hobbit* have each gone for about $200,000—and nobody is very surprised by that. A Shakespeare or Austen or Tolkien first edition is more than the words. It's the brain and the body.

Similarly, it would be nice to believe that we each have some true inner self that has nothing to do with our outer shell—but nothing in nature works that way. The bird that flies, the snake that slithers, and the peacock that spreads its tail are all affected by their physical form. A physical shape gives possibilities as well as limitations. Rolf Pfeifer likes to show this by playing with Legos—specifically the Lego Mindstorms that made a splash when they first came out, allowing kids (and adults) to build and program small robots. Pfeifer once used the set to build four creatures ("if I may call them that," he said) that moved completely differently. One that he referred to as Crazy Bird did wild flips when it was turned on.

"What is the control that underlies the behavior of the system?" he asked provocatively.

The natural assumption is that he programmed it differently—but not so. Legos aren't that complex, and the controls or "brains" of all the creatures was the same. All that changed was how he put the bricks together.

"You have absolutely no way of saying anything about the behavior of the system unless you know how the control is embedded into the physical system and you know the characteristics—shape, morphology, and material characteristics," he said.

Crazy Bird had a rubber piece on one side and a plastic piece on the other. Given the directive to go forward, the difference in friction on the two sides meant that it would spin and flip on its own. Other creatures given the same directions but with wheels that moved straight would respond completely differently. The control may try to prescribe one plan but the physical form also affects what happens. Pfeifer believes that the message applies to human systems too.

"If you only look at the brain and identify neural circuits, you have no way of knowing what these neural circuits mean for the behavior of the organism," he says. "You have to know how the brain is embedded into the physical system."

All animals have embedded smarts. A spider can weave a web and a beaver can build a dam, and we wouldn't expect either of them to do the other's task. Artificial intelligence that exists in silicon chips is unembodied—but researchers have begun to wonder just how much machines learn through their bodies.

At Stanford's Institute for Human-Centered AI, researchers found a new way of looking at how embodiment might affect the development of intelligence. The Institute is led by the estimable Fei-Fei Li, an innovative computer scientist I met when I wrote *The Genius of Women*. Since she has been in the forefront of so many interesting theories about machine learning, I was excited to see that her team developed an elaborate plan to look at "morphological intelligence." Morphology is simply the shape and structure of things—a key to understanding body-mind interactions. They created creatures called unimals (yoo-nimals—universal animals) that could evolve through several generations. As their body shapes changed to face increasingly challenging environments, the unimals' ability to learn new tasks also advanced. After ten generations, the most successful unimal morphologies learned tasks in *half* the time as the original forms.

"We're often so focused on intelligence being a function of the human brain and of neurons specifically," says Fei-Fei Li. "Viewing intelligence as something that is physically embodied is a different paradigm."

It's not just thinking that makes us behave as we do. How we move and experience our environment changes what we learn and know. Pfeifer joins the chorus that says it's time to update our old Cartesian dichotomy of brain and body.

His version: "I act—therefore I am."

The Genius of a Venus Flytrap

Different aspects of body intelligence are being considered and developed around the world. At the esteemed Max Planck Institute in Germany, Metin Sitti leads a group studying physical intelligence—or "intelligence encoded in the body." While we typically rely on our brains and computational intelligence to solve problems, he says that "learning is also possible without any neural computation."

Now wait a minute. Can you really learn without your brain? I immediately thought of a child using fingers (and toes) to do an addition problem, but that's not really physical intelligence. Instead, think about a Venus flytrap. The fly-eating plant uses the sensory hairs on its leaf to generate electrical signals. Sitti says that these effectively make complex decisions including "when to partially close the trap fast to capture the prey, when to close and seal the trap fully, and when to start the digestion process." That's a lot of thinking for one plant—or rather not traditional thinking at all since it's all done by multisensory signals.

Or consider brainless slime mold, which has been taught to solve mazes and puzzles, including the traveling salesman problem. The last one sounds more like the start of joke from a late-night comic than the research of an electrical engineer, so I looked it up. In the traveling salesman problem, you get a list of cities and the distances between them and have to figure out the shortest possible route going from one to the next. Mathematicians and experts in operations research create elaborate programs to figure out the best path, but our one-celled slime mold can solve it too, by leaving a trail of sensors as it searches for food and then learning the energy-efficient shortest routes. No brainpower necessary.

We human beings have the advantage of both neural networks and physical intelligence, and we make a mistake to ignore one for the other. I think we balk a bit at the idea of physical intelligence because it is much more romantic to think of ourselves as mind, soul, and spirit than as a lump of flesh. Aspects of our physical presence seem fixed and unchangeable—whether you are five feet tall or six feet six is not a decision you make, and it's also not one you can change, whereas whether you will study math, linguistics or art history is fully under your control. But that flesh and body, the neurons and sensors and hormones and morphology, are part of who you are. They matter in what you learn and how you experience the world.

We like to think that our mind-power can overcome all odds. And maybe it can. But you are likely to be far happier in life if you follow the lessons of the robots and the Venus flytrap: Recognize the interplay of body and mind and allow that your intelligence exists in every cell of your body.

Stop Telling Children to Sit Down

When my younger son Matt was in fifth grade, I got a call from his teacher, who anxiously said he had something to discuss. Taking a math test that afternoon, Matt had stood up for the whole thirty minutes, leaning over his desk to work.

"How did he do?" I asked.

"He got a 100 percent," the teacher said. "But I wanted to alert you in case there's a stress problem."

I thanked him for letting me know, and that night I told my son about the call and asked for his side of the story.

"I like to stand when I'm doing math. It helps me think better," he said simply.

"Were you nervous about the test?" I asked.

"Not really. I just feel better standing up."

And that was that.

I worried a little bit about the teacher's call, but Matt did fine in school and in real-life too. He wasn't a nervous child, and if he wanted to stretch his legs during a test—well, why should that be a problem? If I had known then about the work of philosopher Andy Clark, I might have had a stronger response to the teacher. I would have explained that instead of being concerned about Matt's technique, he should have everyone else in the class try it too.

Clark, a philosophy professor at the University of Sussex, believes that the you who is you is not just in your brain. It extends to body and environment. Movement informs the way each of us learns and perceives the world, and for many children, if you insist that they sit down, they're not thinking as well. When a kid does something foolish, a scornful teacher or parent might snap, "Use your brain!" But now that researchers are looking at how intelligence and learning have a physical component, the better comeback might be "Use your body!"

Part of thinking for a child is in the body. And it's true for grown-ups too. The gestures you make as you give a speech are part of the thinking process. When a writer or architect uses a notepad or draws pictures on a page, it's not as if they're coming up with an idea and then downloading it onto the page. The writing and drawing are part of the process. As Joan Didion succinctly put it: "I write entirely to find out what I'm thinking." Other authors including Stephen King and Flannery O'Connor made similar comments. It's only in the process of writing that they discover what they want to write.

But this connection between doing and thinking—letting our bodies and actions be part of our learning process—isn't always encouraged. Listening to a lot of podcasts lately, I've heard several comedians and actors talk about the trouble they got into as kids in school—and now it makes sense. Creativity comes from the body as much as the mind, and clever or imaginative children often get labeled disruptive (or worse) when they are simply using all their senses to learn.

Consider a typical school experience. It all starts out okay when children are in pre-K or kindergarten and they are prompted to dig in a sandbox, jump up and down, or make balls out of clay. Teachers understand that the touching, exploring, and creating are part of the learning process. So far so good. But then comes first grade and everything changes as children are expected to sit in chairs and not move for long stretches of time. Even schools that encourage recess (and many more are losing it or cutting back) have kids jammed into their chairs and desks for at least half the day.

All of a sudden, a large number of children—those who learn by doing rather than by listening—become a problem. In preschool they might have been praised for being curious and inquisitive and energetic, but now they are told to sit down and pretend their bodies don't exist. Not all of them can follow that directive, and some 10 percent of children in elementary school get labeled with ADHD—a statistic both shocking and predictable. If we consider 10 percent of children to have a problem, it's probably not the children who have the problem but the system evaluating them.

The proof seems to be in the numbers. When children are younger than five and still encouraged to use their bodies in the interest of learning, barely 2 percent are deemed to have ADHD. Once the rules of school change and they are forced to sit down and not move, the number rises to that stunning 10 percent. It's a good bet that if the children were allowed to use their bodies to express themselves, many of the diagnoses would disappear. One study followed the activity levels of mainstream children aged 10–17 who also had an ADHD diagnosis and found a direct connection between cognitive performance and frequency and intensity of activity. When allowed or encouraged to move, these children think better. They are also more focused, confident, and inventive.

The teachers and parents who anxiously say "Sit down and focus!" have it wrong. The children they're instructing don't want to sit because they instinctively know they *lose* focus that way. Their minds work better when their bodies are moving. Given the setup of most school rooms with rows of desks and overworked teachers, the solution of encouraging movement presents its own challenges. The result is that we fall into the trap of considering a good child to be one who sits still—while the fact is that an even better child might be one who is happily jumping, moving, or wiggling while learning.

Psychologist Lucy Jo Palladino has written extensively about this problem. She points out that if the ADHD diagnosis were bandied about in the late 1800s, Thomas Edison—"a boy who learned only by doing"—would surely have been similarly mislabeled. "At age six, he had to see how fire worked and accidentally burned his father's barn to the ground," she wrote. "The next fall he began school, where he alternated between letting his mind travel to distant places and keeping his body in perpetual motion." Endlessly fidgety, he didn't fit into a classroom and was ultimately homeschooled. I probably don't have to tell you that Edison went on to be one of the greatest inventors of all time. Had he been forced to sit down and constrain body and mind, would he have gone on to originate the electric light bulb, the phonograph, the motion picture camera, and on and on and on?

Palladino refers to those who have this kind of high energy and wandering mind as Edison-trait children—which honestly makes more sense than the medical community's diagnosis. She says that most of them learn better with multisensory experiences, and what works for them might be a better approach for most people. Visual models and bright colors can help learning, as do hands-on experiences. Little children grasp math better if they literally grasp cubes or manipulate an abacus. Older kids understand geometry when they make 3-D projects from the complex shapes they are trying to learn. When my son Matt was in third grade, the clever teacher had the children march around the class with drums and tambourines, making rhythmic songs from their multiplication tables. I can still hear the pulsating chants of

Five times five is twen-ty-five!

Six times six is thir-ty six!

When learning becomes a sensory experience, you start to feel the answers in your bones, not just in your mind. Thinking about it now, I wonder if that early experience of marching math contributed to Matt's natural urge to stand and move when he worked on a problem. There is great joy in being able to feel an answer with your whole body.

What the Magic 8 Ball Says

In TV shows set in offices, people are always doing annoying things—like playing basketball with balled-up pieces of paper tossed into a wastebasket. It happens in real life too. An executive I know keeps a Magic 8 Ball on his desk, which he regularly consults during stressful meetings. I don't think he's actually trying to find out which of the twenty answers will come up when he shakes it, but he likes having something in his hands.

Like my doodles, Magic 8 Balls and crumpled paper balls seem like distractions from focusing on work—but they may simply be the ways we each find to get our bodies involved in the process too. Your body likes to get stirred as much as your mind does, and working at a computer all day isn't its idea of fun. When your body is seeking some sensation, it will be happy to make due with whatever is handy—pulling apart a paper clip or snapping small bits of tape from the roller are desktop favorites. (I've done both.)

When you're fidgeting, your body is looking to find the right level of stimulation. As our workdays get increasingly sedentary, we may need more of those inputs—which could be why fidget toys like spinners began popping up everywhere a few years ago. The schools and offices that ban them for being too distracting are missing the point. Professors Michael Karlesky at New York University and Katherine Isbister at the University of California, Santa Cruz, did a project of "fidget widgets" where they asked people to upload images of the things they like to fiddle with. They pointed out that given the powerful link between hand and brain, fidgeting could enhance creativity, provide focus, and decrease stress. People enthusiastically agreed, offering odes to ballpoint pens to click, Silly Putty to squish, rubber bracelets to snap, and tiny stuffed animals to fondle.

Isbister believes that fidgeting helps children and adults fine-tune the level of stimulation they need. It's a way of managing your attention and emotion through physical activity—and rather than being distracting, the physical motions help you focus. If you find it hard to sit still, the best answer is—don't. Get up and move. Spin a paperclip. Squeeze a fidget widget. For your mind to work its best, your body needs to get involved too. How much movement is necessary varies from person to person. But if you have a high level of embodied self-regulation, don't try to suppress it. Edison didn't.

Dancing In School (and Not Just Dancing School)

I once had the pleasure of meeting the very charming professor and educator Sir Ken Robinson shortly after he spoke to a small group on how to encourage creativity in children. Developing an education system that rewards creativity was one of the great themes of his life, and though I loved chatting with him, I felt slightly dubious about his position. At the time, I wanted to be sure my sons got good grades, aced the SATs, and were prepared for life. He believed that standardized tests were anathema to creativity and we needed to think about intelligence as going well beyond the mind.

Robinson was influential enough to be knighted by Queen Elizabeth, and his TED Talk on how schools kill creativity remains one of the most-viewed of all time. His unorthodox approach eventually won me over too. (Sadly he died in 2020 at age 70.) He worried about an education system created by university professors—he had been one of them—who live in their heads. "They're disembodied. They look at their bodies as…a way to get their head to meetings," he said. Coming from that perspective, they've created a rigid hierarchy of subjects where the arts sit at the very bottom. Robinson saw that as a huge mistake. "I think math is very important, but so is dance. Children dance all the time if they're allowed to—we all do. We all have bodies, don't we? Did I miss a meeting?"

A natural raconteur with a great sense of timing, Robinson made the convincing argument that intelligence is dynamic and interactive. How we learn and what we understand go well beyond the mind. Intelligence is diverse and comes from many sources—many of them physical and involving our bodies. "We think visually, we think in sound, we think kinesthetically.

We think in abstract terms, we think in movement," he said. He thought it unconscionable that as children grow up "we start to educate them progressively from the waist up. And then we focus on their heads. And slightly to one side."

He liked to tell the story of having lunch one day with Gillian Lynne, the great dancer and choreographer. She told him that she had so many problems as a child in school that her mother took her to a psychiatrist when she was eight years old. The mom spent twenty minutes outlining Gillian's shortcomings to the doctor: she was fidgety, disruptive in class, skipping her homework, disturbing people, and not sitting still. The doctor listened to the litany, then took the mom out of the room, leaving Gillian alone with the radio turned on. They watched as she began moving to the music the moment they left. "Gillian isn't sick," he said. "She's a dancer. Take her to a dance school."

Fortunately, the mom followed the good advice, and at the dancing school, Gillian was thrilled to find people just like herself—"people who couldn't sit still, people who had to move to think." She eventually became a soloist at the Royal Ballet, founded her own dance company, and then choreographed dozens and dozens of stage and movie musicals. She teamed up with Andrew Lloyd Weber in creating two of the longest-running shows in Broadway history—*Phantom of the Opera* and *Cats*.

"She's given pleasure to millions and she's a multimillionaire," Robinson said. "Somebody else might have given her medication and told her to calm down."

Robinson got thunderous applause when he used that line in his TED Talk, and yet we continue to give children medication and tell them to calm down. Understanding that we discover and enjoy the world through our bodies changes your perspective. All of us do better and feel happier when movement is encouraged rather than treated as a problem. Intelligence comes in many forms. Fei-Fei Li reveals her genius through computers, and Dame Gillian Lynne conveyed hers through physical expression. Not every fidgety, restless child will end up being Thomas Edison or Gillian Lynne. But we could probably create a lot more of them if we stop insisting that they sit down.

PART SIX

Your Optimistic Body

Your body wants you to be happy, and you can help by understanding the body-mind feedback loops that lead to greater joy. Discover here some very practical techniques, from the clothes you wear to the way you sit, that will improve your outlook every day.

What Language Does Your Body Talk?

> Dance is the hidden language of the soul of the body.
>
> —MARTHA GRAHAM

ON A HOT SUMMER DAY many years ago, I sat in my office at a TV show where I worked knowing that I needed to go to the bank. I tried to imagine waddling the endless distance and decided I couldn't make it. Nine months pregnant, I felt far too exhausted to make the long trek. After I returned from my maternity leave a few months later, I skipped over to the bank one day and started to laugh. The bank was probably three blocks away.

Was the bank close to the TV station or far away? My answer would depend on when you asked. We've all become alarmed lately at the idea of generative AI changing how we perceive reality, but our bodies have been changing our perceptions of the world for a long time. You can't always believe your eyes, because our supposedly objective views can easily be changed by how our bodies feel. They provide new information for our minds to put into the equation. If someone asks you to guess the steepness of a hill in the distance, you probably think you're giving a mathematically based answer. But studies show that your perception depends on whether you're going to be climbing it with a backpack or walking slowly, whether you're in shape and generally healthy or struggling with pain. Distances appear farther and hills look steeper to people who are tired, out of shape, or loaded down with a heavy backpack. It might be that instead of visualizing the actual distance, they defer to their body's calibration of what it will *feel* to walk that distance.

Whenever I go hiking with my husband, he urges me on by saying "We're almost at the top!" He's done it so often that it's become a joke between us. It often seemed to me that we went on forever after his cheery "Almost there!" and I'd grumble about his endless optimism. But the research on bodies and perception has convinced me that he's not really being deceptive. Strong, athletic, and a lover of the outdoors, he truly perceives the mountaintop as a quick sprint away. He can't understand that for us more laggardly types, the distance appears interminable.

You don't see the world and then decide how to act—your physical abilities and action determine how you see the world. When they're scoring well, softball players think the ball is larger than it really is, and basketball players envision the hoop as being larger. Jessica Witt, a psychologist at Colorado State University, wondered if that could work in reverse—changing what the body sees to alter how you perform. She set up a putting green with an optical illusion that could make the hole look bigger—and sure enough, the golfers made better putts.

Our bodies are busy determining how we respond to the world far more than our brain-centric selves could ever imagine. For example, it's possible that your perception of time can change with your heartbeat. In one study I read, a longer heartbeat interval made people think that a sound lasted longer. Another experiment linked your heartbeat to how fearful you may feel. People shown a fearful face at one point in the heart cycle (systole—when the heart is pumping) react more intensely than if shown at another point (diastole—when it's at rest). If this sounds crazy to you, I don't disagree—but I'm excited that these experiments are being done and getting attention. Too often lately, researchers have put people in MRI machines then waved around the brain scans as if they offer conclusive evidence of why we behave as we do. The newer research reminds us that the brain is only one part of the equation. Having bad knees can change how tall a mountain looks to you, and your heartbeat can determine how fearful you feel.

Sarah Garfinkel, a professor of cognitive neuroscience in the UK who led the heartbeat-fear study, says that "there is a growing appreciation that other bodily organs interact with brain function to shape and influence our perceptions, cognitions, and emotions." It all goes back to the theory that philosopher William James proposed in the nineteenth century (and we discussed at the very beginning of this book) that bodily changes lead to emotional feelings. In James's view, if you encounter a bear, your body has an instantaneous response—pounding heart, surging adrenaline, sweaty skin—that leads to the emotion. Your body gets scared and then your mind follows. Some psychologists have stood up against James and argued for the primacy of emotion. They insist your mind can control how you feel, so if you tell yourself not to be afraid when you see the bear you will remain calm. The intellectual wrangling has gone on for more than a century with heated opinions on both sides.

I don't think there will be a quick end to this battle, but new evidence is starting to emerge that James was right—and our bodies play an even *bigger* role in our emotions than previously imagined. In one of the most amazing studies I've encountered recently, Karl Deisseroth, a professor of bioengineering and of psychiatry and behavioral science at Stanford University, developed a process called optogenetics, which modifies the genes of some cells so they can be activated by the flash of a tiny light source. This sounds so wildly like science fiction that I studied the article in the esteemed journal *Nature* and didn't understand a single word of the process. But apparently it worked.

The complicated technique provided a high-tech and non-invasive way to control the heartbeats of mice. Using the technique, a large team of researchers outfitted mice with tiny vests that included a light source capable of changing their heart rate from a baseline of 600 beats per minute to 900. Deisseroth called it a "nice reasonable acceleration" that mimicked what a mouse would experience during a moment of stress or fear.

Okay, so let's clarify. Nothing stressful or fearful actually happened to the mice. The mice didn't see the light source or have any unusual experience. The researchers just raised their heartbeats—mimicking what would happen if something dangerous *had* occurred. The mice responded to the raised heartbeat with what the researchers called "anxiety-like behavior." They limited their exploration of a maze and tried to stay in protected areas. They avoided the center area of a field, just as they would if they were being chased by a predator. When put in a situation that had some real risk (a chance of a mild shock) they became too fearful to seek any reward.

Simply changing the heart rate with no outward reason or threat made the mice anxious or fearful. Looking at the findings, the researchers concluded that "both the body and brain must be considered together to understand the origins of emotional or affective states." It's unlikely that we can do a similar experiment with humans any time soon, but we can still extrapolate what happens. Your body has a physical response to any source (an intruder in your kitchen or an inexplicable change in your heart rate from a light), and then your mind follows with an interpretation. *My heart is racing so I'm scared!* Your behavior changes to fit how your body is feeling.

The light study makes it clear that the brain monitors internal states to figure out the correct behavioral response. Your body reacts, your mind interprets, your behavior follows. With such an erudite and distinguished intellect as William James, one can't really imagine him wagging his finger and saying "I told you so." But with these findings, he would have had every right.

Next up might be looking at other bodily responses. Deisseroth's optogenetics could be used to investigate additional ways that changing the body changes the emotions. If someone's shoulders are made to tense, does that make them feel more anxious? How about if you could induce butterflies in the stomach? Would that make you assume you're nervous? We don't really have to wait for more studies—because all the evidence we have says your body is pretty powerful at telling you how to feel. You can use that information right now to make yourself feel happier and calmer. Do your own conscious body scanning if you're feeling anxious or nervous. Shake out your shoulders to relax the muscles, or take a deep breath to calm a pounding heart. Use your body to help send the positive, happy messages you want your brain to hear. You might be surprised at how well it works.

The Power of a Silk Nightie

The happy-making information your body feeds the brain comes from a wide range of sources and senses. I recently started seeing a lot of posts on social media about "dopamine dressing"—how you can change what you wear to get a feel-good boost. Usually it involves bright colors instead of the neutrals that most of us wear, but the connection between mood and clothes goes well beyond putting on a yellow sweater. The texture of the cloth against your body changes your sensations, and the way a fabric drapes can make you feel happy or despairing.

Optogenetics requires a fancy lab, but you can test out how clothing changes your mood by going into your closet or the nearest department store. I asked two young women to join me in the fitting rooms at Saks for our own little experiment. The deal was they would put on the items but not look in the mirror, offering instead a word or two about how the item made them feel. We started with a body-hugging silk chemise that elicited sighs of pleasure from both and the descriptor "sensuous" from one and "sexy" from the other. Very soft jersey leggings got a double vote of "cozy." The most expensive item, a floral silk crepe de chine caftan with a price tag over $2,000, did not get the response the fancy designer might have hoped. One of the women said she felt awkward and the other said that despite the soft material "I feel frumpy."

Then they tried on the same items but looked in the mirror. Without even realizing it, they now commented on how they looked rather than how they felt. One worried that the previously sexy chemise wasn't flattering to her chest, and the other complained that the cozy leggings accentuated her hips. Both liked the caftan better in the mirror than they had before, noting that it was eye-catching and would send a message of being creative and upscale.

"I can see how someone rich would buy this to show off but feel ridiculous wearing it," one of them said.

The caftan, flannel, and chemise went back on the rack, but to thank them for their time (and feel like I hadn't exploited Saks), I bought each of them a soft cotton T-shirt that they said made them feel "relaxed" and "happy."

As we left the store, we talked about how surprising it was that their reactions had been so different when they judged how the clothes felt on their bodies rather than how they looked in the mirror. The lesson seemed to be that the clothes we wear serve a double purpose: They send a message to other people but also to ourselves.

This clothing double-whammy shows up in unexpected ways. Clothes can affect your mood and your confidence enough to change how you function in the world. In one study that always surprised me, women who tried on a bathing suit in private before a math test did worse than women who had tried on a sweater. For whatever reason, seeing themselves in a bathing suit changed their self-perception enough to undermine their abilities on a test.

In another study about a dozen years ago, researchers had

What Language Does Your Body Talk?

half the volunteers put on a white lab coat, which they hypothesized would send the wearer a message of professionalism and care. They gave them a test that measured ability to pay attention to detail—and reported that those wearing the white coat made half as many errors on one part of the test as the volunteers who stayed in street clothes. For an additional part of the experiment, they told some of the people that the white coat was a doctor's lab coat and others that it was a painter's smock. The attention abilities increased only when the coat was associated with a doctor.

It was a cool finding and one that makes intrinsic sense. Clothes help announce who we are to other people, but they also send a physical message to our own brains. The researchers even gave their finding the smug title "enclothed cognition"—a clever play on the embodied cognition we have been talking about where body influences mind. The study was widely cited by other academics and reported with great exuberance in popular magazines and websites. Alas, another team of researchers couldn't replicate the findings, and the original duo agreed that this "cast doubt" on their findings. But they bounced back—pointing out that enough other studies have shown the power of clothes to influence our mind (though not with attention tests) that "the core principle of enclothed cognition—what we wear can influence how we think, feel, and act—is generally valid."

Okay, I'm willing to go with that. Put on the uniform of a policeman, nurse, painter, or astronaut and you immediately send a message to your brain about how to behave. Men who put on a suit for a negotiation did dramatically better in one study than those asked to wear casual clothes. The suit wearers gained some $2 million in profits during the negotiation, while those dressed in sweatpants and sandals did only about a third as well. Wearing the suit induced dominance and even resulted in higher testosterone levels.

The power of clothes can vary for different people and situations. The Silicon Valley guys who follow Mark Zuckerberg in wearing hoodies as a uniform probably negotiate better that way than they would in suits. I often work from home, but I've been known to change my outfit a couple of times in the morning before I sit down at my computer. Nobody will see me—but I need to feel right to myself to get anything done.

"Clothes speak to others, but they have the power to speak to us too," says UK psychologist Karen Pine. "Sensations and associations in the body lead to new ideas in the mind."

Pine has found that women tend to put on jeans and baggy tops when they're feeling depressed, perhaps as a way of disappearing from the world and disassociating from their own bodies. When happy, you're more likely to put on well-cut, figure-enhancing clothes that have pretty fabrics. But how about if, when you're feeling glum, you make yourself put on the happy clothes? The effect could trigger a positive feedback loop from body to brain that lifts your mood. "The brain's functions are primed by sights, smells, and experiences," says

Pine. "Now we know our mental processes can also be primed by a piece of clothing."

I recently found an online site that makes great eyeglasses very inexpensively, so I now have a wardrobe of glasses that includes pairs in red, orange, and pink. I was wearing the orange glasses the other day, and a woman I didn't know stopped me in the grocery store and said, "Excuse me, those are very happy glasses." I like having happy glasses. Spreading a little joy immediately makes you feel happier too.

Clothes and accessories affect our emotions and become the guardians of them too. What we wear creates feelings in us, and we also transfer our feelings to our clothes. The cute polka-dot skirt I happened to be wearing the day I got hired for an important job remains one of my favorites, and the dark blue dress I wore the day my dad died never came out of my closet again. It sounds like we're attributing power to clothes that they don't really have—that skirt had nothing to do with my getting hired (at least I don't think so)—but clothes create a visceral feeling in our bodies. Items that become good-luck symbols for you actually can bring positive vibes. You may stand slightly differently when you wear them and hold your head higher. When your body feels more confident, your whole demeanor changes.

You don't have to dress like Carrie Bradshaw in order to feel happier. Just be aware of how you face the world differently depending on what you're wearing. Recognizing the power of enclothed cognition could change what you buy. If that silk chemise makes you feel sensuous and powerful, you should probably go for it and avoid the awkward-making caftan, no matter how in style it might be. Your clothes make an impression on other people—but first they make an impression on you.

Your Body Memory

You never forget how to ride a bike, the saying goes, and it's pretty much true. Get on a two-wheeler after years away, and you may have a few wobbles, but you'll probably be pedaling securely in no time. Trying to think through each step of what to do (*climb on seat, put feet on pedals...*) will likely be counterproductive. You need to rely on your muscle and nerve pathways to remember the motions that were previously engrained. The same holds true for skiing and swimming and tossing a baseball and practicing many other learned physical skills. Long after you've forgotten the name of your first tennis teacher, your body still knows how to coordinate the motions of a good forehand.

People often refer to this as muscle memory, but motor memory is probably more accurate. The sequence of movements you need for biking or swimming or even tying your shoelaces gets embedded in neural pathways of the central nervous system. Some researchers think the actions get embedded in several *different* pathways, which could explain why they come back so readily. You don't "remember" how to ride a bike in the same way that you remember your first boyfriend or the square root of one hundred. Instead, you reactivate one of the pathways. When you want to bike or swim or ski again, your body knows exactly what to do.

"Clearly, there's a huge difference in the way that motor memories are formed," says Jun Ding, a neurologist at Stanford who has done some breakthrough research on the subject. He suggests that you can think of motor memory like a highway system. "If Interstate 101 and Highway 280 are both closed, you could still get to Stanford from San Francisco," he says. It's a fairly local reference, but you get the point. Motor memory may be unique in that your body has different ways to access it.

Some new research suggests that actual muscle memory may also exist. If you managed to get to the gym often to lift weights when you were in your twenties, don't despair if you've given it up for a while. The resistance training creates certain alterations on a cellular level to your muscle fibers. The cells gain extra nuclei, which remain for months or years after you stop. Even as the muscle mass begins to dwindle, the cellular changes linger. When you start lifting weights again, your strength will return much faster than it did initially or than it would if you'd never previously exercised. It's as if the muscles are saying *We know how to do this!*

When it comes to physical skills, our bodies often do best when they are allowed to function without any conscious interference. Once your body has laid down the neural pathways for an activity, thinking about it will just get in the way. I realized this one day when I was driving in the car with a chatty and mechanically curious seven-year-old in the backseat.

"Do you have your foot on the gas?" he asked. When I said yes, he said, "Is the gas pedal on the left side or the right side?" I had to think about it, and after I answered he said, "Okay, so then the brake is on the left and you have to brake now, don't you?" I did need to stop, but I suddenly felt very uncomfortable about where to put my foot. My body knew how to go from gas to brake, but consciously thinking about the movement was hindrance, not help. (If you want to try this as an experiment yourself next time you're out driving, I recommend a quiet street.) When your body knows what to do, your thinking brain barging in only gets it confused. The same is true if you're playing the guitar or keyboard and decide to think about where to put your fingers for each chord or even if you consciously focus on where to put each foot as you walk down the street.

I once heard a story about a rock climber who had a specific sequence of steps she followed each time she prepared to climb. She would tighten the belts on the harness, tie into the rope, clip into the belay device.… She had done the long process so many times that it had become a deeply engrained pathway. One day she stopped in the middle of the sequence to tie her shoes. When she started again, her long-established pattern had been disrupted, and without realizing it, she missed one of the crucial steps for clipping into a rope. She suffered a dangerous fall later and blamed herself for the disrupted preparation.

What Language Does Your Body Talk?

It seemed as if her body knew that there were a certain number of things to do in preparation, and tying her shoes became one of them. When her physical rhythm was thrown off balance, her mind got the necessary steps confused.

I've used this lesson in the most mundane way. When I'm leaving my apartment, I usually have a bunch of things to take with me like computer, phone, Kindle, keys, and wallet. The recycling container is just down the hall, but I always make a point of taking out the recycling first and then coming back. Too many times when I've taken the old magazines to toss on my way, I get to the elevator and realize that I've forgotten one of the items I need. It's as if my body knows how many objects it should be carrying and the recycling becomes one of them. Our body memory has its own reason.

The delicate balance of body and mind became apparent to the world at the summer Olympics in 2021 when the great gymnast Simon Biles stunned everyone by dropping out of the competition. Biles had thirty-two World and Olympic medals on her shelves—more than anyone, ever—and four gymnastics elements named after her. She never lost, and she had been expected to sweep the gold medals again. But after an unexpectedly bad vault where she did one and a half flips instead of the planned two and a half, Biles realized her body wasn't performing right. Her muscle memory refused to kick in. When that happens, the now-panicked conscious mind steps in and tries to control the action. But it simply can't. Biles later described it as having the "twisties"—getting confused in the air about the position of your body where you can't tell up from down. "It's the craziest feeling ever. Not having an inch of control over your body," she wrote on Instagram. Your mind doesn't know how to make you do two and a half flips over a vault. You need the deeply entrenched neural pathways in your body, the motor memory, to be in control.

Her coaches tried various techniques to help and get her back to at least one of the event finals. "Every avenue we tried, my body was like, *Simone, chill. Sit down. We're not doing it.* And I've never experienced that," she said later. What did it feel like? Biles said to imagine that you've had good eyesight your whole life then wake up one morning and can't see anything. People tell you to keep going as if nothing happened. "You'd be lost, wouldn't you? That's the only thing I can relate it to."

It's hard to know exactly what interfered with Biles's perfectly trained body doing its perfectly learned routines. Perhaps stress upset the balance. She had to deal with the huge expectations placed on her as well as the strangeness of the Olympics, which had no spectators because of the ongoing pandemic. Her parents couldn't be with her, and she'd also been under enormous pressure in previous months as the sexual abuse of gymnasts by a repellent doctor came to light. Whatever the cause, she was wise enough to put pride aside and understand that she couldn't go on.

"It's honestly petrifying trying to do a skill but not having your mind and body in sync," she said shortly after she withdrew. "Ten out of ten do not recommend."

What Language Does Your Body Talk?

We ask a lot of our bodies, and we don't always appreciate how amazing it is when they perform—whether driving a car, riding a bike, or doing a double-twisting double flip dismount from a balance beam (called the Biles, of course). Biles eventually recovered and started winning championships again a couple of years later. We live with our bodies every day, but sometimes it feels like we are clueless about how they function. Neurobiologist Dean Buonomano says that in his own field, "It remains to be seen if the human brain can understand the human brain." Interestingly, it requires our bodies to understand our brains. And both function at the highest level only when they are working together.

17

The Body-Mind Happiness Plan

> I must learn to be content with being happier than I deserve.
>
> —JANE AUSTEN

I WOKE UP THE OTHER night and looked at the digital clock, which said 2:22, and my whole body immediately responded with pleasure. Some loud noise outside had awakened me in the middle of the night, but I felt a visceral surge of happiness. A gratitude moment!

The symmetry of the numbers lining up on a clock only happens a few times any day or night. If by chance I look at the clock at the precise right moment, like 3:33 or 5:55, I have that engrained response of delight. The pattern began when I hosted a podcast about gratitude. On one episode, I suggested finding little signals that could remind you to shake off negative thoughts and focus on something positive. A listener contacted me with the idea of using a digital clock. He said that if he happens to see a clock when all the numbers are aligned, he takes it as a sign to think about something happy. It's completely unplanned and unfiltered and doesn't even occur all that often. But those magical moments start to become a way to focus on the good of the moment.

I liked the concept so much that it became part of my own habits, ultimately so engrained in my neural pathways that when I saw the clock in the middle of the night, it was like a chime in my head. *Happiness moment!*

You can engrain your own moments of positivity by connecting a physical action to a time to be happy. You could use brushing your teeth as a gratitude reminder. Take those two minutes when you're brushing in the morning and at night to reframe the day's events in a positive way. It makes tooth brushing a lot more interesting.

Or perhaps decide that getting in the car to drive from the grocery store will be a signal to focus only on positive thoughts. The bad events of the day (or the fact that the store was out of your favorite ice cream) can wait. Connecting positive thoughts to a physical experience helps make a happiness habit a lot easier to instill. Your body wants you to be happy. Give it the chance to let good feelings flow and they will.

Ode to Joy

We all have embarrassing moments from childhood, and mine involves the first time my body and mind refused to work together. It happened in fourth grade when I worked very hard to prepare for my very first piano recital. My mom came early on the big day. She sat in the front row and expected me to do well.

At my turn to perform, I sat down on the piano bench, and as I put my fingers on the keys, my hands began trembling. I had no idea what was happening. How could I play with quivering fingers? I feebly hit one note then immediately bungled the next one. My two hands wouldn't coordinate. Stunned, I stopped and tried again. But my hands were shaking harder now, and I produced a sound so dissonant I began to cry.

Ruining the recital proved more than mortifying—it upended my understanding of myself. I had always been resolute and hard-working and thought I could do anything I set my mind to. But as I sat at the piano, my thinking mind seemed to have no sway. Much as I tried, I couldn't tell myself to simply play the song. My body had its own ideas. The bafflement I felt at finding my body outside my conscious control haunted me for years, and I have clearly never forgotten the feeling.

Back then, I didn't know about adrenaline or cortisol or stress responses. I didn't know that sometimes our bodies respond to a situation first and our conscious minds hurry to catch up. Perhaps if I had understood the rush of hormones and neurotransmitters taking place as I sat at the piano, it wouldn't have been so terrifying. I would have known that my body sensed danger and had gone on high alert. In many situations, the response might have saved my life. For playing "Ode to Joy" on the piano, though, it couldn't have been more wrong.

Understanding what is happening in our bodies and the messages they are sending can transform our attitude, mood, and performance. By understanding your body's signals, you can help your mind interpret them differently—so you feel happier and more joyously integrated with body and mind. Too often we inhabit our bodies but understand amazingly little about how they function. We end up being at odds with our bodies rather than feeling at home in them.

Whether you're playing on a piano or in a park, writing a novel or an email, getting a sandwich or getting a new job, you do better and feel happier if body and mind are working together. What we really want is the opposite of what I experienced at the recital—to feel that we are one unit of body and mind, moving forward in joyous synchronicity.

Hopefully much of what we've talked about in this book can help you get more in sync with yourself. If you want to increase the possibility of your body making you happy, take a week to use some of the findings we've uncovered. By integrating actions into your daily life, you can help yourself find a new level of body-mind connection and, with it, a new level of happiness.

Here is a seven-step plan to get you started. Pick any week to do it—or spread it out over a longer stretch. Either way, you'll feel better and happier when you're done. Better yet, you can keep repeating the plan until it's part of your everyday life.

Monday: Create a Cozy Environment

The physical experiences of your body get reinterpreted by your brain—which can be more literal than you might expect. When you hold a warm cup of coffee you typically judge the people you meet afterward as being warmer. Sit in a soft chair and you feel more giving toward others than when you're in a hardback seat. The time-honored idea of solving problems with a hot bath and a warm beverage is not so far off.

To feel happier, make sure your body receives the sensations that send a message of general well-being. Surround yourself with soothing objects that bring physical pleasure—a steaming cup of coffee, a soft fuzzy sweater, a warm fireplace. Denmark, regularly listed as one of the happiest countries in the world, promotes a culture of hygge, which is simply a coziness in your environment that brings contentment. While there's no direct translation, hygge derives from a term related to the English world for "hug." Wherever you live, you can make your environment feel like a warm hug.

To get started today, add something sensually satisfying to your space—a velvet pillow, a soft throw, some colorful dishes, or anything that gives you a physical sense of pleasure. Then let yourself take a moment to savor them. Share the happy space with other people, and you can spread the good feeling. Being connected with friends and sharing objects that send a message of warmth is a surefire way to let your brain find new pathways to happiness. Make sure your neurons are sending a message of happiness and well-being that both body and brain can understand.

The clothes you wear become part of that personal environment you create, so instead of considering what's fashionable at the moment, find the items that put you in a positive mood. The color, structure, and feeling of fabric against skin can all change how you feel. The sensations your body gets from your clothes get processed in your mind—and that feedback loop can lift your mood or bring it down. Whether you're going out or staying inside, make sure that what you wear sends the happy-making message you want your brain to hear.

Tuesday: Let Your Whole Body Be Happy

Your mind responds strongly to the sensory experiences your body receives from outside, and it also interprets what's going on inside, getting information from all the muscles and neurons. Today, no matter how you're feeling, take a moment to make yourself smile. Your facial expression doesn't just indicate how you feel—it can *create* how you feel. Your brain reads the stimuli from the muscles to make a prediction about your emotional state.

Your smile or frown is really just the start. When you're feeling down or depressed you tend to hunch your shoulders and slump forward. Your body is literally downward-facing. Today, make a point to catch yourself in any moments of depressed posture—then stand or sit up straight and make your eyes look ahead or upward. Since this is the physical stance you take when you're feeling confident or victorious, your brain reacts to it and begins processing a more positive story.

If you spend a lot of time curled over a laptop or iPad, start a new habit today of pausing a couple of times an hour to stretch your arms upward. Let the neural fibers going from body to brain send a flow of positive feeling. When you're staring at a screen and concentrating hard, there's a good chance you've also tightened the muscles of your forehead. When you're stretching your arms, also unclench your facial muscles by lifting your eyebrows upward. Then put your lips into a small smile, and the facial feedback will kick in to tell your brain that all is well—or at least better than you might think.

Wednesday: Take Time Outside

Taking time to be in nature improves your ability to focus and can even strengthen your memory. You gain a sense of connectedness that makes you feel more grateful about the world. In one study, people sitting in a natural setting like a park improved their well-being more than people sitting in a man-made setting like a parking lot. You'd guess that one, since who wouldn't rather be in a park than a parking lot? But the reason goes beyond aesthetics. Being in nature actually changes your brain physiology. Even watching videos of nature scenes can improve your emotional state.

Since being outside makes you feel happier, take a few minutes today to go someplace where your mind can relax and make its own connections. The scenes and sounds of nature offer a soft fascination—interesting enough to provide constant sensory stimulation but not so overwhelming that you can't think of anything else. The sensory experience leads to physical changes that ultimately lower stress and improve health and well-being.

If you live in an urban area, you can find simple ways of connecting with nature like visiting a rooftop garden or buying flowers to bring home. Simply wandering through a park or along a river can change your mood, and if you can get to someplace that includes water, the change to your happiness might be even greater. Standing at the edge of a riverbank, you are engaged in an ever-changing scene. The movement of the water, the light reflecting off its surface, the changing sounds provide a visceral experience that registers in your body to make you feel happier.

While you're outside, pause a few moments to look upward and take in the sky. Part of the pleasure of nature is the enchantment of ephemeral phenomena—the scenes that occur once and never again. The drifting clouds, twinkling stars, and glowing rainbow are always changing, but the movement occurs at a leisurely speed that won't cause your body to respond with alarms of danger. Your sensory experience increases and so does your happiness.

Thursday: Reinterpret Your Body's Signals

Reframing a negative situation is one of the great gifts you can give yourself. It puts you in control of events rather than being wafted about by them. You recognize that there can be good and bad in any situation, and you are the one who gets to decide where to focus. Finding the positive doesn't remove the difficulties, but it puts them in a different frame. *I'm frustrated that the electricity went out in the storm, but the candles look pretty and the fireplace is cozy!* You choose which part of the event you want to make the centerpiece of your thoughts.

Usually the new perspective takes place entirely in our minds, but you'll succeed much better by first acknowledging what your body is feeling. When an outside event causes a change in your physical state, your mind tries to figure out how to interpret it. Pounding heart and rapid breathing? Hmmm… you could be scared, excited, or in love. Your unconscious mind takes a shot at deciphering, but that doesn't have to be the last word. You can help yourself by providing a different perspective.

Today you're going to recognize that the same physical signals can mean many different things. If you're light-headed and breathing hard before a presentation, don't assume you're scared and likely to do a terrible job. Instead, flip that view. Recognize the adrenaline-induced state of high arousal—but attribute it to an eagerness to show your abilities. Emotions are your brain's prediction of what your bodily sensations mean, and it's perfectly all right if you help the brain to recalibrate.

Turning a negative guess into a positive one can transform the experience you're about to have.

Even in the most distressing moments of our lives, we need a way to move forward. Reframing to find a bit of light in the situation can be just the inspiration you need.

Friday: Don't Exercise—Just Move

When you're in motion, your body releases endorphins and neurotransmitters that improve your mood and raise your spirits. The chemical anandamide, which gets released during vigorous exercise, is very similar to THC, the active ingredient in cannabis. But you don't need extreme levels of exertion to improve your mood. Simply being in motion by taking a walk or tending your garden can improve your life satisfaction. Find an activity that you enjoy, and think of it as a pleasure you give yourself, not something you "have" to do.

The real pleasure of exercise comes when movement involves a coordination of body and mind—your body taking pleasure in motion while your mind experiences new surroundings. While your body moves, you also get the sensory stimulation that changes your mindset and improves focus and energy. The movement helps reduce stress and opens you to more creative thinking.

Today, plan to go someplace new where you can simply play. Do something you enjoy and that feels natural to you. Jog freely along a hiking trail, dance in your backyard, or dare to swim in the ocean (even if it's a little cold). When your body is healthy and strong and functioning at peak form, your mind gets the message that all is well. Whatever your current state of mind, moving will improve your mood and lessen any impending depression. You can think of it as a new law of thermodynamics: A body in motion stays happy.

Saturday: Eat for Pleasure

Your body sends information to your brain on the nutrients that it needs, but it's sometimes hard to hear the signals amid the loud racket of diet advice, marketing, and faux food. Today you're going to separate the signal from the noise by paying attention to what makes your body satisfied. Instead of thinking about willpower and calories, you'll focus on the sensory pleasure of eating. Start to think about healthy foods like apples or tomatoes in terms of their taste and satisfaction rather than the idea that they're good for you. You'll enjoy them much more.

Remember that your greatest delight with any food comes in the first few bites. Your taste sensations get a new thrill, and your brain registers the pleasure of something new. But keep eating too much and the neural pathways get dulled, and both body and mind stop paying attention. Overeating actually diminishes your pleasure in the food.

Children seem to naturally understand eating for pleasure. They'll cheer as the cupcakes appear at a birthday party and eagerly sit down to take the first bite of frosting. But after a few bites, they've had enough. The thrill of the new treat is gone, and they jump up, ready for the next activity. The parents clear away the half-eaten cupcakes, usually in some amazement.

To take a lesson from the birthday cupcakes, think about a food that actually gives you joy. How much of it do you need to increase your happiness—and at what point does overindulgence send your happiness spiraling downward? Satisfy your sensual desire for the food and when that is sated, know that you can stop.

Even in this age of plentiful food, we have a drive to indulge when we can. We unconsciously worry about what's ahead and so eat too much of what's currently available. Today, develop a work-around to this instinct. You might keep a small piece of chocolate in a drawer for a late-night indulgence or eat half your dessert at dinner and save the rest for later. Knowing that you have some sensuous pleasure ahead relieves your mind of its anxiety about future scarcity and allows your body to call the shots.

Not everything about your weight or metabolism is under your control, but instead of going on endless diets, adopt the Happy Body Lifestyle. Stick to as much fresh, real food as possible. Avoid overly processed foods, which confuse your body and undermine your mood and well-being. Pay attention to what your body is saying, and let food be a source of controlled pleasure rather than guilt.

Sunday: Walk Your Troubles Away

The celebrated poet William Wordsworth took long walks every day through the English countryside, strolling some twenty miles each day. He often headed out on even more challenging hikes, including one foot journey that took him through France, Germany, and the Alps. Wordsworth traversed some one hundred and eighty thousand miles in his lifetime, and he composed his greatest poems on his rambles. He once estimated that nine-tenths of his verses had "been poured out in the open air." The combination of physical movement and the sensuous appeals of nature unleashed his creative joy and his poetic inspirations.

Wordsworth wasn't alone—many of the greatest philosophers and writers in history have found a powerful connection between moving their body and unleashing new ideas. "There is something in walking that stimulates and enlivens my ideas," said philosopher Jean Jacques Rousseau whose greatest essays were inspired by his long daily rambles. "I can hardly think at all if I stay still. My body has to be on the move to set my mind going." Rousseau believed he could get close to his soul and examine the meaning of life and death only while he was in motion.

We think of the high-flying and esoteric ideas of philosophers like Nietzsche and Kant as coming from their minds, but it's fascinating to discover that their ideas were grounded in physical connections. Nietzsche went on long, steep hikes and famously said not to believe in any idea "that was not born in the open air and of free movement—in which the muscles do not also revel." He described being overcome by the intense feelings released in him as he trekked along the landscapes, discovering several times that "during my long walks I had wept too much, and not sentimental tears but tears of happiness."

Frédéric Gros, now a professor of philosophy in Paris, has taken up the mantle of walking as a way to find fundamental joy. He says that the opportunities for happiness are like "gold threads in the world's fabric. They ought to be seized." Whether you are walking in the city or the country, on vigorous hikes or leisurely ambles, those golden threads of happiness become more visible. You breathe, you move, you feel alive and aware.

So today, find an hour (or seven or eight, if you prefer) to go for a walk. You don't have to think about anything in particular or make any plans. Your mind will wander as your feet connect to the ground. In walking, your body finds a lightness and happiness that drift to your mind. You exist in the moment, body and mind connected and part of the greater world. From that connection often arises a lightness to both heart and thought. You are in sync with yourself and ready to do anything.

Body and mind working together in a loop of hope, inspiration, gratitude, and new ideas. Achieving that joyous connection may be the ultimate source of happiness.

Suggested Reading

John Bargh, *Before You Know It: The Unconscious Reasons We Do What We Do*, Touchstone, 2017

Lisa Feldman Barrett, *How Emotions Are Made: The Secret Life of the Brain*, Mariner Books, 2017

Antonio Damasio, *The Feeling of What Happens: Body and Emotion in the Making of Consciousness*, Harcourt Brace and Co., 1999

David Eagleman, *Incognito: The Secret Lives of the Brain*, Vintage, 2012

Daniel Gilbert, *Stumbling on Happiness*, Albert A. Knopf, 2006

Peter Godfrey-Smith, *Other Minds: The Octopus, the Sea, and the Deep Origins of Consciousness*, Farrar, Straus, and Giroux, 2016

Frédéric Gros, *A Philosophy of Walking* (translated by John Howe), Verso reprint, 2015

Hans Rocha IJzerman, *Heartwarming: How Our Inner Thermostat Made Us Human*, W. W. Norton and Company, 2021

Alan Jasanoff, *The Biological Mind: How Brain, Body, and Environment Collaborate to Make Us Who We Are*, Basic Books, 2018

Roman Krznaric, *Carpe Diem Regained: The Vanishing Art of Seizing the Day*, Unbound, 2018

George Lakoff and Mark Johnson, *Metaphors We Live By*, University of Chicago Press, 1980

Daniel E. Lieberman, *Exercised: Why Something We Never Evolved to Do Is Healthy and Rewarding*, Vintage, 2020

Wallace J. Nichols, *Blue Mind: The Surprising Science That Shows How Being near, in, on, or under Water Can Make You Happier, Healthier, More Connected, and Better at What You Do*, Little Brown and Company, 2014

Annie Murphy Paul, *The Extended Mind: The Power of Thinking Outside the Brain*, Houghton Mifflin Harcourt, 2021

Charles Spence, *Gastrophysics: The New Science of Eating*, Viking, 2017

Acknowledgments

I love writing books and am grateful to the many readers who let me keep doing this. Thanks to Alice Martell, who cared about this topic from the very first moment and helped guide me through many early versions. Her wisdom and tenacity are unmatched. Warm thanks also to Erin McClary for her early support and confidence in the project and to wonderful editor Kate Roddy for picking up the mantle and making everything work better. I knew the book was in good hands the first time I spoke to her. I am thrilled to be working with the powerful team at Sourcebooks including Kayleigh George and Liz Kelsch who are always bursting with great ideas. My admiration to Dominique Raccah for building such an impressive publishing house.

Many doctors, psychologists, neuroscientists and researchers graciously spent time sharing their work and insights with me, and they have my deepest appreciation. Their wisdom infuses every page. Extra thanks to Dr. Sean Mackey at Stanford and Dr. Tor Wager at Dartmouth who are doing incredible work in pain management and generously explained the latest findings and techniques to me, and to John Bargh whose fascinating research inspired my interest in the whole field of embodied cognition.

I wrote much of this book at the Yale Club where powerhouse librarian Caroline Bartels was a wonderful help, always finding the book or article that I couldn't. My dear friend Robert Masello keeps me laughing and our regular phone calls make every day brighter. I am grateful to have many friends who provide encouragement and perspective with notable thanks to Susan Fine, Lynn Schnurnberger, Lisa Dell, Beth Schermer, Stephen Landsman, Lorna Schofield, Mike Berland, Jean Hanff Korelitz, Leslie Berman, Candy Gould, and the members of my NYC book group, and great admiration to new friend and educator Mimosa Jones Tunney.

My amazing sons Zach and Matt make it easy to write about happiness because they are smart, funny, perceptive, and kind and bring me joy every day. Annie and Pauline are simply the best daughters-in-law in the world with unsurpassed energy, charm, and graciousness. My husband Ron always gets mentioned last but he is my first reader and first advisor, my terrific friend and dear love who makes everything possible. We have written many chapters together and I am excited for those yet to come.

Notes

Introduction

Comic John Mulaney: John Mulaney probably wasn't the first person to joke about the role of the body. Thomas Edison reportedly once that "The chief function of the body is to carry the brain around."

"Without our bodies there can be no consciousness": Antonio Damasio, *The Feeling of What Happens: Body and Emotion in the Making of Consciousness* (New York: Harcourt Brace and Co., 1999), 3.

"Your body is part of your mind": Kim Armstrong, "Interoception: How We Understand Our Body's Inner Sensations," *Observer*, Association for Psychological Science (September 25, 2019), https://www.psychologicalscience.org/observer/interoception-how-we-understand-our-bodys-inner-sensations

"The body's captive audience": Damasio, *The Feeling of What Happens*, 150.

Chapter 1: What Your Apple Watch Can't Tell You

New research by Jordan Etkin: Jordan Etkin, "The Hidden Cost of Personal Quantification," *Journal of Consumer Research* 42, no. 6 (April 2016): 967–984, https://doi.org/10.1093/jcr/ucv095; "Why Counting Your Steps Could Make You Unhappier," Duke Fuqua School of Business, December 21, 2015, https://www.fuqua.duke.edu/duke-fuqua-insights/etkin-counting-steps.

"Our feeling body": Judson Brewer, phone interview with the author, August 30, 2022. (All Brewer quotes are from this interview.)

The weight of the clipboard: Joshua M. Ackerman, Christopher C. Nocera, and John A. Bargh, "Incidental Haptic Sensations Influence Social

Judgments and Interactions," *Science* 328, no. 5986 (June 2010): 1712–1715, https://doi.org/10.1126/science.1189993.

"Our brains take their input": "Philosophy in the Flesh: A Talk with George Lakoff," Edge, March 1999, https://edge.org/3rd_culture/lakoff/lakoff_p1.html.

"We cannot think just anything": "Philosophy in the Flesh."

"First impressions are likely to be influenced": Ackerman, "Incidental Haptic Sensations."

"Your consciousness is like a tiny stowaway": David Eagleman, *Incognito: The Secret Lives of the Brain* (New York: Vintage Books, 2012), 98.

those who had been primed with iced coffee: Lawrence E. Williams and John A. Bargh, "Experiencing Physical Warmth Promotes Interpersonal Warmth," *Science* 322, no. 5901 (October 2008): 606–607, https://doi.org/10.1126/science.1162548.

"It's fascinating": John Bargh, interview with the author, September 13, 2022. (All Bargh quotes are from this interview.)

included in the game: Chen-Bo Zhong and Geoffrey Leonardelli, "Cold and Lonely: Does Social Exclusion Literally Feel Cold?" *Psychological Science* 19, no. 9 (September 2008): 838–842, https://doi.org/10.1111/j.1467-9280.2008.02165.x.

actual body temperature went down: Hans Ijzerman et al., "Cold-blooded Loneliness: Social Exclusion Leads to Lower Skin Temperatures," *Acta Psychologica* 140, no. 3 (July 2012): 283–288, https://doi.org/10.1016/j.actpsy.2012.05.002.

a study at UCLA hospital: Tristan K. Inagaki et al., "A Pilot Study Examining Physical and Social Warmth: Higher (Non-Febrile) Oral Temperature Is Associated with Greater Feelings of Social Connection," *PLOS One* 10, no. 2 (June 2016), https://doi.org/10.1371/journal.pone.0156873.

Chapter 2: How Your Body Makes You Happy

"When there's a change in your body": Erik Peper, phone interview with the author, October 4, 2022. (All Peper quotes are from this interview.)

"tower over your circumstances": Maya Angelou, *Rainbow in the Clouds: The Wisdom and Spirit of Maya Angelou* (New York: Random House, 2014), 58.

"They use cars and planes to transport their brains": Kara Swisher, "Sex Bots, Religion, and the Wild World of A.I.," November 1, 2020, *Sway*, produced by Paula Szuchman, podcast, 38:40, nytimes.com/2021/11/01/opinion/sway-kara-swisher-jeanette-winterson.html.

"brains have always developed in the context of a body": Katrin Weigmann, "Does Intelligence Require a Body," *EMBO Reports* 13, no. 12 (December 2012): 1066–1069, https://doi.org/10.1038/embor.2012.170.

he wrote one of the original papers: Murad Alam et al., "Botulinum Toxin and the Facial Feedback Hypothesis: Can Looking Better Make You Feel Happier?" *Journal of the American Academy of Dermatology* 58, no. 6 (June 2008), 1061–1072, https://doi.org/10.1016/j.jaad.2007.10.649.

"It's a real possibility": Kenneth Arndt, phone interview with the author, October 18, 2022.

"I have often observed": Edmund Burke, *A Philosophical Inquiry into the Sublime and Beautiful*, rev. ed. (repr., Oxford, UK: Oxford University, 2015), 164.

A much-reported study: Fritz Strack, Leonard Martin, and Sabine Stepper, "Inhibiting and Facilitating Conditions of the Human Smile: A Nonobtrusive Test of the Facial Feedback Hypothesis," *Journal of Personality and Social Psychology* 54, no. 5 (1988): 768–777, https://doi.org/10.1037/0022-3514.54.5.768.

The study hit some controversy: E. J. Wagenmakers et al., "Registered Replication Report: Strack, Martin, and Stepper (1988)," *Perspectives on Psychological Science* 11, no. 6 (October 2016), https://doi.org/10.1177/1745691616674458.

But not so: Tom Noah, Yaaciv Schul and Ruth Mayo, "When Both the Original Study and Its Failed Replication Are Correct: Feeling Observed Eliminates the Facial-Feedback Effect," *Journal of Personality and Social Psychology* 1114, no. 5 (2018): 657–664, https://doi.org/10.1037/pspa0000121.

Collecting data from 138 studies: Christopher Paley, "Smiling Does Make You Happier-Under Carefully Controlled Conditions," *Guardian*, October 17, 2018, https://www.theguardian.com/science/2018/oct/17/smiling-does-make-you-happier-under-carefully-controlled-conditions.

Nicholas Coles helped organize: Nicholas A. Coles et al., "A Multi-Lab Test of the Facial Feedback Hypothesis by the Many Smiles Collaboration," *Nature Human Behavior* 6 (October 2022): 1731–1742, https://doi.org/10.1038/s41562-022-01458-9.

even when people: Nicholas Cole et al., "Fact or Artifact? Demand Characteristics and Participants' Beliefs Can Moderate, But Do Not Fully Account for, The Effects of Facial Feedback on Emotional Experience," *Journal of Personality and Social Psychology* 124, no. 2 (2023): 287–310, https://doi.org/10.1037/pspa0000316.

He asked people to imitate angry expressions: Andreas Hennenlotter et al., "The Link Between Facial Feedback and Neural Activity Within Central

Circuitries of Emotion: New Insights From Botulinum Toxin-Induced Denervation of Frown Muscles," *Cerebral Cortex* 19, no. 3 (March 2009): 537–542, https://doi.org/10.1093/cercor/bhn104.

standing up to scrutiny: Jara Schulze et al., "Botulinum Toxin for the Management of Depression: An Updated Review of the Evidence and Meta-Analysis," *Journal of Psychiatric Research* 135 (March 2021): 332–340, https://doi.org/10.1016/j.jpsychires.2021.01.016.

Chapter 3: Your Mixed-Up Mind

One study out of Iowa University: Brad J. Bushman, "Does Venting Anger Feed or Extinguish the Flame? Catharsis, Rumination, Distraction, Anger, and Aggressive Responding," *Personality and Social Psychology Bulletin*, 28, no. 6 (2002): 724–731, https://doi.org/10.1177/0146167202289002.

One psychologist: The scoffing connection between Freud and "I'm a Little Teapot" is not meant to be taken literally. Freud died in 1939, the year the song came out. But the psychologist is right to be mocking about the catharsis theory. A very full discussion of the many ways Freud has led us astray can be found in Frederick Crews, *Freud: The Making of an Illusion* (New York: Metropolitan Books, 2017).

"venting to reduce anger": Brad Bushman, phone interview with the author, November 4, 2022. (All Bushman quotes are from this interview.)

"All emotions use the body": Damasio, *The Feeling of What Happens*, 52.

In one famous study: Donald Dutton and Arthur Aron, "Some Evidence for Heightened Sexual Attraction Under Conditions of High Anxiety," *Journal of Personality and Social Psychology* 30, no. 4 (1974): 510–517, https://doi.org/10.1037/h0037031.

"Emotions are your body's best guess": Lisa Feldman-Barrett, "We Don't Understand How Emotions Work. A Neuroscientist Explains Why We Often Get It Wrong," BBC Science Focus, October 28, 2021, Sciencefocus.com/the-human-body/what-are-emotions.

In a string of experiments: Alison Wood Brooks, "Get Excited: Reappraising Pre-Performance Anxiety as Excitement," *Journal of Experimental Psychology* 143, no. 3 (June 2014): 1144–1158, https://10.1037/a0035325.

In one study: Jeremy P. Jamieson et al., "Turning the Knots in Your Stomach into Bows: Reappraising Arousal Improves Performance on the GRE," *Journal of Experimental Social Psychology* 46, no. 1 (January 2010): 208–212, https://doi.org/10.1016/j.jesp.2009.08.015.

did dramatically better on the practice GRE: The people who were told that anxiety could improve their performance scored an average of 738 on the math portion of the practice test versus 683 for the controls; on the actual test, they scored 770 versus 705. Interestingly, the reappraisal had no effect on verbal scores. The researchers conjectured that a math problem requires higher executive functioning, which is most affected by reappraisal. It may also be that most people are more anxious when faced with a math test, so the reappraisal has more dramatic effects.

"People have come to believe that stress is bad": Jeremy Jamieson, phone interview with the author, November 17, 2022. (All Jamieson quotes are from this interview.)

Chapter 4: How Your Senses Give You Joy

In a study done: Vera Morhenn, Laura Beavin, and Paul Zak, "Massage Increases Oxytocin and Reduces Adrenocorticotropin Hormone in Humans," *Alternative Therapies in Health and Medicine* 18, no. 6 (November 2012), https://pubmed.ncbi.nlm.nih.gov/23251939/.

Everything that has form: Johann Gottfried Herder and Jason Geiger, *Sculpture: Some Observations on Shape and Form from Pygmalion's Creative Dream* (repr., Chicago: University of Chicago Press, 2002), 33.

people valued the project far higher: Michael Norton, Daniel Mochon, and Dan Ariely, "The IKEA Effect: When Labor Leads to Love" *Journal of Consumer Psychology* 22, no. 3 (July 2012): https://doi.org/10.1016/j.jcps.2011.08.002.

if you see what I mean: Neil Gaiman, *The Graveyard Book* (New York: Harper Collins, 2008), chap. 4.

Chapter 5: Why Blue and Green Are the Happiest Colors

less likely to become nearsighted: Justin Sherwin et al., "The Association Between Time Spent Outdoors and Myopia in Children and Adolescents: A Systematic Review and Meta-Analysis," Ophthalmology 119, no. 10 (October 2012): 2141–2151, https://10.1016/j.ophtha.2012.04.020.

"substantially happier outdoors": George MacKerron, interview with author, January 18, 2023; George MacKerron and Susana Mourato, "Happiness Is Greater in Natural Environments," *Global Environmental Change* 223, no. 5 (October 2012): 992–1000, https://doi.org/10.1016/j.gloenvcha.2013.03.010; George MacKerron, "Mappiness: How Space & Time Impact Our Well-Being," speech, World Government Summit, 2019.

Notes

Trees rather than a brick wall: RS Ulrich, "View Through a Window May Influence Recovery from Surgery," *Science* 224, no. 4647 (April 1984): 420–421, https://doi.org/10.1111/j.1467-9280.2008.02225.x.

people did better on memory tests: Marc G. Berman, John Jonides, and Stephen Kaplan, "The Cognitive Benefits of Interacting with Nature," *Psychological Science* 19, no. 12 (December 2008): https://doi.org/10.1111/j.1467-9280.2008.02225.x.

When you are engaged: Marc Berman, phone interview with author, October 2014.

nature is the best medicine: Kathryn Schertz and Marc Berman, "Understanding Nature and Its Personal Benefits," *Current Directions in Psychological Science* 28, no. 5 (2018): https://journals.sagepub.com/doi/10.1177/0963721419854100.

Heart rate and blood pressure go down when you're outdoors: Marcia P. Jiminez et al., "Associations Between Nature Exposure and Health: A Review of the Evidence," *International Journal of Environmental Research and Public Health* 18, no. 9 (May 2021): https://doi.org/10.1016/j.ufug.2020.126932.

being in natural environments reduces cortisol levels: Wenfei Yao, Xiaofeng Zhang, and Qi Gong, "The Effect of Exposure to the Natural Environment on Stress Reduction: A Meta-Analysis," *Urban Forestry and Urban Greening* 57 (January 2021): https://doi.org/10.1016/j.ufug.2020.126932.

A study done: Mark E. Beecher et al., "Sunshine on My Shoulders: Weather, Pollution, and Emotional Distress," *Journal of Affective Disorders* 205 (November 15 2016): 234–238, https://doi.org/10.1016/j.jad.2016.07.021.

"We tried to take into account": Jon McBride, "Sunshine Matters a Lot to Mental Health," BYU Physics and Astronomy, November 2016, Physics.byu.edu/department/news/2016-11-sunshine-matters-a-lot-to-mental-health.

could change your likelihood of being admitted: Uri Simonsohn, "Clouds Make Nerds Look Good: Field Evidence of the Impact of Incidental Factors on Decision Making," SSRN (August 2006): http://dx.doi.org/10.2139/ssrn.906872.

daily stock returns: David Hirshleifer and Tyler Shumway, "Good Day Sunshine: Stock Returns and the Weather" (Working paper no. 2001–3, Dice Center): http://dx.doi.org/10.2139/ssrn.265674.

forest rangers in a part of Iceland: Trevor Nace, "Iceland Forest Service Recommends Hugging Trees Since You Can't Hug People," *Forbes*, April 14, 2020, https://www.forbes.com/sites/trevornace/2020/04/14/icelandic

-forest-service-recommends-hugging-trees-since-you-cant-hug-people/?sh=6f3c4f99784c.

"flowers are mood elevators": Lewis Miller, *Flower Flash* (New York: Monacelli Press, 2021), 167.

strong positive effects: Jeannette Haviland-Jones, "An Environmental Approach to Positive Emotion: Flowers," *Evolutionary Psychology* 3, no. 1 (2005), https://doi.org/10.1177/147470490500300.

"why is nature so good?": Mathew White, Zoom interview with the author, January 23, 2023. (All White quotes are from this interview.)

"bubbly hot springs": Jancee Dunn, "What Rainn Wilson Learned Searching for Joy Around the World," *New York Times*, May 19, 2023, https://www.nytimes.com/2023/05/19/well/mind/rainn-wilson-happiness.html.

Chapter 6: Places That Make Your Spirits Soar

"quantifiable consequences for our well-being": Chanuki Illushka Seresinhe, Tobias Preis, and Helen Susannah Moat, "Quantifying the Impact of Scenic Environments on Health," *Scientific Reports* 5 (November 2015): https://doi.org/10.1038/srep16899.

"hardwired to understand fractals": Florence Williams, "Why Fractals Are So Soothing," *Atlantic*, Jantuary 2017, https://www.theatlantic.com/science/archive/2017/01/why-fractals-are-so-soothing/514520/.

"redundant rather than simple": Robert Venturi, *Complexity and Contradiction in Architecture* (1966 repr., New York: Museum of Modern Art, 1977), 16.

Mathias Basner, a doctor, researcher, and sleep expert: Mathias Basner et al., "Auditory and Non-Auditory Effects of Noise on Health," *Lancet* 383, no. 9925 (April 2014): 1325–1332, https://10.1016/S0140–6736(13)61613-X.

to the damaging effects of noise: Mathia Basner and Sarah McGuire, "WHO Environmental Noise Guidelines for the European Region: A Systematic Review on Environmental Noise and Effects on Sleep," *International Journal for Environmental Research and Public Health* 15, no. 3 (March 2018): 1–45, https://10.3390/ijerph15030519.

The results were stunning enough: Arline L. Bronzaft and Dennis P. McCarthy, "The Effect of Elevated Train Noise on Reading Ability," *Environmental Behavior* 7, no. 4 (December 1975): https://doi.org/10.1177/001391657500700.

even while we're sleeping: Mathias Basner, "Why Noise Is Bad for Your Health and What You Can Do About It," filmed2018, TEDMED video, 10:22, https://www.tedmed.com/talks/show?id=730074.

the breaks proved to be far more relaxing: Luciano Bernardi, Camillo Porta, and P. Sleight, "Cardiovascular, Cerebrovascular, and Respiratory Changes Induced by Different Types of Music in Musicians and Non-Musicians: The Importance of Silence," *Heart* (British Cardiac Society) 92, no. 4 (April 2006): 445–452, https://10.1136/hrt.2005.064600.

who listened to the sound of birds: Emil Stobbe et al., "Birdsongs Alleviate Anxiety and Paranoia in Healthy Participants," *Scientific Reports* 12 (October 2022): https://doi.org/10.1038/s41598-022-20841-0; Ryan Hammoud et al., "Smartphone-Based Ecological Momentary Assessment Reveals Mental Health Benefits of Birdlife," *Scientific Reports* 12 (October 2022): https://doi.org/10.1038/s41598-022-20207-6.

Chapter 7: Why Wine Tastes Better in Paris

Our senses are talking: Charles Spence, Zoom interview with the author, September 21, 2023. (All Spence quotes are from this interview.)

wine tastes better: Hilke Plassman et al., "Marketing Actions Can Modulate Neural Representations of Experienced Pleasantness," *PNAS* 105, no. 3 (January 2008): 1050–1054, https://doi.org/10.1073/pnas.0706929105.

flimsy cutlery: Charles Spence, *Gastrophysics: The New Science of Eating* (New York: Viking, 2017), 97.

"is also its setting" Spence, *Gastrophysics*, 9.

"We eat with our eyes, ears": Spence, *Gastrophysics*, forward by Heston Blumenthal, xi.

tallest skyscraper in midtown Manhattan: The high-in-the-sky observation deck visited was the Summit at One Vanderbilt.

an alief: Tamar Szabo Gendler, "Alief and Belief," *Journal of Philosophy* 105, no. 10 (October 2008): 634–663, https://doi.org/10.5840/jphil20081051025.

mind-body discordance: Paul Rozin and April Fallon, "A Perspective on Disgust," *Psychological Review* 94, no. 1 (1987): 23–41, https://doi.org/10.1037/0033-295X.94.1.23.

Chapter 8: What Body Positivity Really Means

Robin Williams famously quoted: The poem being quoted is the opening lines of "To the Virgins, To Make Much of Time," by Robert Herrick, written in the mid-1600s.

talking about a stranger: Emily Pronin, Christopher Y. Olivola, and Kathleen A. Kennedy, "Doing Unto Future Selves As You Would Do Unto Others: Psychological Distance and Decision Making," *Personality and Social Psychology Bulletin* 34, no. 2 (February 2008): https://doi.org/10.1177/0146167207310.

Harvard psychologist Dan Gilbert: Dan Gilbert, *Stumbling on Happiness* (New York: Knopf, 2006), 1.

"existential crime of the century": Roman Krznaric, *Carpe Diem Regained: The Vanishing Art of Seizing the Day* (London: Unbound, 2017), 4.

"flicking through the channels": Krznaric, *Carpe Diem*, 62.

"a long war against pleasure": Fiona Macdonald, "What It Really Means to 'Seize the Day,'" BBC Culture, May 2017, https://www.bbc.com/culture/article/20170517-what-it-really-means-to-seize-the-day.

talking together in coffee shops: Sidney Jourard, "An Exploratory Study of Body Accessibility," *British Journal of Clinical Psychology* 5 (1966): 221–231, https://doi.org/10.1111/j.2044-8260.1966.tb00978.x.

the philosopher Johann Gottfried von Herden: Herden and Gaiger, *Sculpture*, introduction.

texture or pressure or shape: Ewa Jarocka, J. Andrew Pruszynski, and Roland S. Johansson, "Human Touch Receptors Are Sensitive to Spatial Details on the Scale of Single Fingerprint Ridges," *Journal of Neuroscience* 41, no. 16 (April 2021): 3622–3634, https://doi.org/10.1523/JNEUROSCI.1716-20.2021.

knowledge of the outside world: Herden and Gaiger, *Sculpture*, introduction.

"animal-assisted therapy": Marin Hedin, "Therapy Dogs May Unlock Health Benefits for Patients in Hospital ICU," Johns Hopkins HUB, February 2018.

"sex is good for us": Nicole Cirino, interview with the author, February 6, 2023. (All Cirino quotes are from this interview.)

half a dozen Marvel flicks: Eliana Dockterman, "Why Aren't Movies Sexy Anymore?" *Time*, February 3, 2023, https://time.com/6251447/sexy-movies-magic-mike-3/.

Chapter 9: The Happy Body Food Plan

"in a modern world": Mark Bittman and David L. Katz, *How to Eat: All Your Food and Diet Questions Answered* (New York: Harvest, 2020), 6.

"Balance is good": Bittman and Katz, *How to Eat*, 12.

"The only way to change": Judson Brewer, phone interview with the author, August 30, 2022. (All Brewer quotes are from this interview.)

"Sensory pleasure peaks": Pierre Chandon, Zoom interview with the author, June 6, 2023. (All Chandon quotes are from this interview.)

In one groundbreaking study: Alia J. Crum, William R. Corbin, Kelly D. Brownell, Peter Salovey, "Mind Over Milkshakes: Mindsets, Not Just Nutrients, Determine Ghrelin Response," *Health Psychology* 30, no. 4 (July 2011): 424–429, 10.1037/a0023467.

mild depression and anxious days: Eric M. Hecht et al., "Cross-Sectional Examination of Ultra-Processed Food Consumption and Adverse Mental Health Symptoms," *Public Health Nutrition* 25, no. 11 (November 2020): 3225–3234, https://doi.org/10.1017/S1368980022001586.

devoid of nutrients: "7 Worst Snacks Your Dietitian Would Never Eat," Cleveland Clinic Health Essentials, December 29, 2020, https://health.clevelandclinic.org/7-worst-snacks-dietitian-never-eat/.

"Feed your body what it needs": Ruth Reichl, "Constant Craving," *Allure*, April 2014, https://www.allure.com/story/food-critic-ruth-reichl-on-weight-management.

half as many depressive symptoms: Sarah E. Jackson et al., "Is There a Relationship Between Chocolate Consumption and Symptoms of Depression? A Cross-Sectional Survey of 13,626 US Adults," *Focus On: Depression, Suicidality and Health* 36, no. 10 (October 2019): 987–995, https://doi.org/10.1002/da.22950.

to truly change a mood: Charles Spence, "Comfort Food: A Review," *International Journal of Gastronomy and Food Science* 9 (October 2017): 105–109, https://doi.org/10.1016/j.ijgfs.2017.07.001.

Chapter 10: How Exercise Makes You Happy

sitting uses eighty calories per hour: Robert Schmerling, "The Truth Behind Standing Desks," Harvard Health Blog, September 2016, https://www.health.harvard.edu/blog/the-truth-behind-standing-desks-2016092310264; Seth Creasy et al., "Energy Expenditure During Acute Periods of Sitting, Standing, and Walking," *Journal of Physical Activity and Health* 13, no. 6 (June 2016): 573–578, https://doi.org/10.1123/jpah.2015-0419.

"people who study and promote exercise": Lieberman, *Exercised*, 47.

"crosses the blood-brain barrier": Amby Burfoot, "Runner's High," *Runner's World*, April 2004, https://www.runnersworld.com/runners-stories/a20866434/runners-high-0/.

One large survey: Supa Pengpid and Karl Peltzer, "Sedentary Behaviour,

Physical Activity and Life Satisfaction, Happiness and Perceived Health Status in University Students from 24 Countries," *International Journal of Environmental Research and Public Health* 16, no. 12 (June 2019): https://doi.org/10.3390/ijerph16122084.

life satisfaction and happiness improved: Hsin-Yu An et al., "The Relationships between Physical Activity and Life Satisfaction and Happiness among Young, Middle-Aged, and Older Adults," *International Journal of Environmental Research and Public Health* 17, no. 13 (July 2020): https://doi.org/10.3390/ijerph17134817.

send your mood in the opposite direction: Pamela Wicker and Bernd Frick, "The Relationship between Intensity and Duration of Physical Activity and Subjective Well-Being" *European Journal of Public Health* 25, no. 5 (October 2015): 868–872, https://doi.org/10.1093/eurpub/ckv131.

Physical activity improves well-being: Andreas Heissel et al.; "Exercise as Medicine for Depressive Symptoms? A Systematic Review and Meta-analysis with Meta-regression," *British Journal of Sports Medicine* 57, no. 16 (August 2023): https://doi.org/10.1136/bjsports-2022-106282; Felipe Schuch et al.; "Physical Activity and Incident Depression: A Meta-Analysis of Prospective Cohort Studies;" *American Journal of Psychiatry* 175, no. 7 (July 2018): 631–648, https://doi.org/10.1176/appi.ajp.2018.17111194; Ben Singh et al.; "Effectiveness of Physical Activity Interventions for Improving Depression, Anxiety and Distress: An Overview of Systematic Reviews," *British Journal of Sports Medicine* (February 2023): https://doi.org/10.1136/bjsports-2022-106195.

lower rates of depression: Lynette Craft and Frank Perna, "The Benefits of Exercise for the Clinically Depressed," *Primary Care Companion to the Journal of Clinical Psychiatry* 6, no. 3 (2004): 104–111, https://doi.org/10.4088/pcc.v06n0301.

Chapter 11: Everybody Hurts (Sometimes)

Nearly 65 million Americans: "Chronic Back Pain," Georgetown University, McCourt School of Public Policy report, https://hpi.georgetown.edu/backpain/.

an X-ray or MRI scan: TS Carey and J. Garrett, "Patterns of Ordering Diagnostic Tests for Patients with Acute Low Back Pain: The North Carolina Back Pain Project," *Annals of Internal Medicine* 125, no. 10 (November 1996): 807–814, https://doi.org/10.7326/0003-4819-125-10-199611150-00004.

having disc problems: MC Jensen et al., "Magnetic Resonance Imaging of the Lumbar Spine in People Without Back Pain," *New England Journal of Medicine* 331, no. 2 (July 1994): 69–73, https://doi.org/10.1056/NEJM199407143310201.

"trends toward worse outcomes": Roger Chou et al., "Imaging Strategies for Low-back Pain: Systematic Review and Meta-analysis," *Lancet* 373, no. 9662 (February 2009): 463–472, https://doi.org/10.1016/S0140-6736(09)60172-0.

"Pain is an experience": Sean Mackey, Zoom interview with the author, March 7, 2023. (All Mackey quotes are from this interview.)

specific cortical fields: Yueqing Ping et al., "Sweet and Bitter Taste in the Brain of Awake Behaving Animals," *Nature* 527 (November 2015): 512–515, https://doi.org/10.1038/nature15763.

"taste is really in your brain": Simon Makin, "A Matter of Taste: Can a Sweet Tooth Be Switched Off in the Brain?", *Scientific American*, May 30, 2018, https://www.scientificamerican.com/article/a-matter-of-taste-can-a-sweet-tooth-be-switched-off-in-the-brain/.

"abnormal amplification within the nervous system": Helen Ouyang, "Can Virtual Reality Help Ease Chronic Pain?" *New York Times*, April 26, 2022, https://www.nytimes.com/2022/04/26/magazine/virtual-reality-chronic-pain.html.

"our brain can ramp pain up": Tor Wager, Zoom interview with the author, March 22, 2023. (All Wager quotes are from this interview.)

Chapter 12: Pain, Pain, Go Away

"it's still unreal to me": "Western Carolina University," Western Carolina University, 2012, YouTube video, 1:31, https://www.youtube.com/watch?v=E-mLj9h3MRQ&t=29s..

"I didn't say anything,": Lauren Campbell, "Bruins' Jake DeBrusk Details Pain He Played Through in Winter Classic," *New England Sports Network*, February 2023, https://nesn.com/2023/02/bruins-jake-debrusk-details-hand-lower-body-injuries/.

for at least a year: Yoni Ashar et al., "Effect of Pain Reprocessing Therapy vs. Placebo and Usual Care for Patients with Chronic Back Pain: A Randomized Clinical Trial," *JAMA Psychiatry* 79, no. 1 (2022): 13–23, https://doi.org/10.1001/jamapsychiatry.2021.2669.

only about two percent of people: Jason M. Waton, David L. Strayer, "Supertaskers: Profiles in Extraordinary Multitasking Ability," *Psychonomic Bulletin and Review* 17, no. 4 (August 2010): 479–485, https://doi.org/10.3758/PBR.17.4.479.

Music changes how people perceive pain: David Bradshaw et al., "Effects of Music Engagement on Responses to Painful Stimulation," *Clinical Journal of Pain* 28, no. 5 (June 2012): 418–27. https://doi.org/10.1097/AJP.0b013e318236c8ca.

by diverting your attention: Sigrid Lunde et al., "Music-induced Analgesia: How Does Music Relieve Pain?" *Pain* 160, no. 5 (May 2019): 989–993, https://doi.org/10.1097/j.pain.0000000000001452.

aromatherapy can be effective: Shaheen Lakhan, Heather Sheafer, and Deborah Tepper, "The Effectiveness of Aromatherapy in Reducing Pain: A Systematic Review and Meta-Analysis," *Pain Research and Treatment* 16 (December 2016): https://doi.org/10.1155/2016/8158693.

feeding their own pain experience: Laura Petrini and Lars Arendt-Nielsen, "Understanding Pain Catastrophizing: Putting Pieces Together," *Frontiers in Psychology* 11 (December 2020):https://doi.org/10.3389/fpsyg.2020.603420; Madelon Peters, Johan Vlaeyen, and Annemarie Kunnen, "Is Pain-related Fear a Predictor of Somatosensory Hypervigilance in Chronic Low Back Pain Patients?" *Behaviour Research and Therapy* 40, no. 1 (January 2002): 85–103, https://doi.org/10.1016/s0005-7967(01)00005-5.

"There appear to be few data": "Pathways to Prevention Workshop: The Role of Opioids in the Treatment of Chronic Pain," speech, National Institutes of Health, September 29–30, 2014, Executive Summary, https://prevention.nih.gov/programs-events/pathways-to-prevention/workshops/opioids-chronic-pain/workshop-resources#final report.

first proposed the idea: Ronald Melzack and Patrick Wall, "Pain Mechanisms: A New Theory," *Science* 150, no. 3699 (November 1965): 971–979, https://doi.org/10.1126/science.150.3699.9; Joel Katz and Brittany N Rosenbloom, "The Golden Anniversary of Melzack and Wall's Gate Control Theory of Pain: Celebrating 50 Years of Pain Research and Management," *Pain Research and Management* 20, no. 6 (November–December 2015): 285–286, https://doi.org/10.1155/2015/865487.

Chapter 13: Sugar Pills Are Sweeter Than You Think

long Latin name: The eye health supplement contained Astaxanthin, a naturally occurring carotenoid derived from the microalgae Haematococcus pluvialis.

More than half of: Jenny Jia, Natalia A. Cameron, Jeffery A. Linder, "Multivitamins and Supplements Benign Prevention or Potentially

Harmful Distraction?" *JAMA* 327, no. 23 (2023): 2294–2295, https://doi.org/10.1001/jama.2022.9167.

An analysis of research: Eliseo Guallar et al., "Enough Is Enough: Stop Wasting Money on Vitamin and Mineral Supplements," *Annals of Internal Medicine* 159, no. 12 (December 2013): 850–851, https://doi.org/10.7326/0003-4819-159-12-201312170-00011.

increased with some supplements: Goran Bjelakovic and Christian Gluud, "Surviving Antioxidant Supplements," *Journal of the National Cancer Institute* 99, no. 10 (May 2007): 742–743, https://doi.org/10.1093/jnci/djk211.

Major medical journals: Guallar, "Enough Is Enough"; Jia, Cameron, and Linder, "Multivitamins and Supplements."

mind-body self-healing processes: Ted Kaptchuk et al., "Placebos Without Deception: A Randomized Controlled Trial in Irritable Bowel Syndrome," *PLOS One* (December 22, 2010): https://doi.org/10.1371/journal.pone.0015591.

and other ailments: Ted Kaptchuk and Franklin Miller, "Open Label Placebo: Can Honestly Prescribed Placebos Evoke Meaningful Therapeutic Benefits?" *BMJ* 363 (Oct 2018): https://doi.org/10.1136/bmj.k3889.

82 percent of the benefit: Irving Kirsch, "Antidepressants and the Placebo Effect," *Zeitschrift Fur Pscyhologie* 223, no. 3 (2014): 128–134, https://doi.org/10.1027/2151-2604/a000176; Irving Kirsch, Thomas Moore, Alan Scoboria, and Sarah Nicholls, "The Emperor's New Drugs: An Analysis of Antidepressant Medication Data Submitted to the U.S. Food and Drug Administration," *Prevention & Treatment* 5, no. 1 (July 2002): https://doi.org/10.1037/1522-3736.5.1.523a; Irving Kirsch et al., "Initial Severity and Antidepressant Benefits: A Meta-Analysis of Data Submitted to the Food and Drug Administration," *PLOS Medicine* 5, no. 2 (February 2008), https://doi.org/10.1371/journal.pmed.0050045.

"I call them placebos": Kirsch, "Antidepressants," *PLOS Medicine* 5, no. 2 (February 2008), https://doi.org/10.1371/journal.pmed.0050045.

no better than the placebo treatment: J. Bruce Moseley, "A Controlled Trial of Arthroscopic Surgery for Osteoarthritis of the Knee," *New England Journal of Medicine* 347 (July 2002): 81–88, https://doi.org/10.1056/NEJMoa013259.

absolutely no change: Adrian Sander, Corrine Glesne, and James Bodfish, "Conditioned Placebo Dose Reduction: A New Treatment in Attention-Deficit Hyperactivity Disorder?", *Journal of Developmental & Behavioral Pediatrics* 31, no. 5 (June 2010): 369–375, https://doi.org/10.1097/DBP.0b013e3181e121ed.

drugs get the credit: Daniel Waschbusch et al., "Are There Placebo Effects in the Medication Treatment of Children with Attention-Deficit Hyperactivity

Disorder?", *Journal of Developmental and Behavioral Pediatrics* 30, no. 2 (April 2009), 158–168, DOI: https://doi.org/10.1097/DBP.0b013e31819f1c15.

Chapter 14: The Neuroscience of Invention

Oppezzo even titled her research findings: Marily Oppezzo, Daniel L. Schwartz, "Give Your Ideas Some Legs: The Positive Effect of Walking on Creative Thinking," *Journal of Experimental Psychology Learning, Memory, and Cognition* 40, no. 4 (July 2014): 1142–1152, https://doi.org/10.1037/a0036577.

the moment my legs begin to move: Laurence Stapleton, *H.D. Thoreau: A Writer's Journal* (New York: Dover Publications, 2011), 64.

would be to stop running: Marily Oppezzo, "What to Be More Creative? Go For a Walk," filmed 2017, in Stanford, CA, TEDxStanford video, 5: 16, https://www.ted.com/talks/marily_oppezzo_want_to_be_more_creative_go_for_a_walk/transcript.

"I run by myself": Ryan Holiday, "Malcolm Gladwell on Running, Writing, and Storytelling," June 19, 2021, in *Daily Stoic* , https://dailystoic.com/malcolm-gladwell/.

the directions their hands point: In the painting School of Athens, Plato is pointing upwards while Aristotle holds his hand outstretched and flat. They are possibly discussing their views of the world, with Plato believing in the importance of spirituality while Aristotle emphasized the grounded and empirical.

prescribed path: Supriya Murali and Barbara Handel, "Motor Restrictions Impair Divergent Thinking During Walking and During Sitting" *Psychological Research* 86 (January 2022): 2144–2157, https://doi.org/10.1007/s00426-021-01636-w.

without external constraints: Murali, "Motor Restrictions."

mental flexibility: Michael Slepian and Nalini Ambady, "Fluid Movement and Creativity," *Journal of Experimental Psychology* 141, no. 4 (November 2012): 625–629, https://doi.org/10.1037/a0027395.

modern dance encourages improvising: Andreas Fink and Silke Woschnjak, "Creativity and Personality in Professional Dancers," *Personality and Individual Differences* 51, no. 6 (October 2011): 754–758, https://doi.org/10.1016/j.paid.2011.06.024.

"spread throughout our bodies": Alan Jasanoff, *The Biological Mind: How Brain, Body and Environment Collaborate to Make Us Who We Are* (New York: Basic Books, 2018), chap. 2, introduction.

learned them underwater: D. R. Godden and A. D. Baddeley,

"Context-Dependent Memory in Two Natural Environments: On Land and Underwater," *British Journal of Psychology* 66, no. 3 (1975): 325–331, https://doi.org/10.1111/j.2044-8295.1975.tb01468.x.

50 percent of people: Adam Grant, "Goodbye to MBTI, the Fad That Won't Die," *Psychology Today*, September 2013, https://www.psychologytoday.com/us/blog/give-and-take/201309/goodbye-to-mbti-the-fad-that-wont-die.

does not exist: The Myers-Briggs test was originally created based on personality distinctions that were outlined by the Swiss psychiatrist Carl Jung, who thought that binary distinctions were a helpful way of thinking about people. But most traits exist on a continuum, and even he would be shocked by the certainty that the test pretends to offer. He once wrote that "there is no such thing as a pure extravert or a pure introvert. Such a man would be in the lunatic asylum."

Chapter 15: How Your Body Makes You Smart

Doodling had helped them: Jackie Andrade, "What Does Doodling Do?" *Applied Cognitive Psychology* 24, no. 1 (January 2010): 100–106, https://doi.org/10.1002/acp.1561.

their ability to solve problems: Girija Kaimal et al., "Functional Near-Infrared Spectroscopy Assessment of Reward Perception Based on Visual Self-Expression: Coloring, Doodling, and Free Drawing," *Arts in Psychotherapy* 55 (September 2017): 85–92, https://doi.org/10.1016/j.aip.2017.05.004.

"to acquire its own experiences," Katrin Weigmann, "Does Intelligence Require a Body?", EMBO Reports 13, no. 12 (November 2012): 1066–1069, https://doi.org/10.1038/embor.2012.170.

"an apple is all these things": Weigmann, "Intelligence."

distributed throughout the organism: Rolf Pfeifer and Josh Bongard, *How the Body Shapes the Way We Think: A New View of Intelligence* (Boston: MIT Press, 2007), 20.

embedded into the physical system: Rolf Pfeifer, "How the Body Shapes the Way We Think," filmed 2013, in Zurich, Switzerland, TEDx video, 19:30, https://www.youtube.com/watch?v=mhWwoaoxIyc.

different paradigm: Katharine Miller, "How Bodies get Smarts: Simulating the Evolution of Embodied Intelligence," Stanford University Human-Centered Artificial Intelligence Newsletter, October 2021.

learn without the brain: Mettin Sitti, "Physical Intelligence as a New Paradigm," *Extreme Mechanics Letters* 46 (July 2021): https://doi.org/10.1016/j.eml.2021.101340.

"find out what I'm thinking": Joan Didion, "Why I Write," *New York Times Book Review* (December 5, 1976), https://www.nytimes.com/1976/12/05/archives/why-i-write-why-i-write.html.

let kids move: Lucy Jo Palladino, *Dreamers, Discoverers & Dynamos: How to Help the Child Who Is Bright, Bored and Having Problems in School* (New York: Ballantine Books 1999), 3.

give them fidgets: Michael Karlesky and Katherine Isbister, "Fidget Widgets Project," https://fidgetwidgets.tumblr.com/.

fine-tune the level of stimulation: Katherine Isbister, "Popping Toys, the Latest Fidget Craze, Might Reduce Stress For Adults and Children Alike," *The Conversation*, May 7, 2021, https://theconversation.com/popping-toys-the-latest-fidget-craze-might-reduce-stress-for-adults-and-children-alike-158746.

"They're disembodied": Ken Robinson, "Do Schools Kill Creativity?" filmed June 26, 2006, in Monterey, CA, TED video, 19:12, https://www.ted.com/talks/sir_ken_robinson_do_schools_kill_creativity?language=en. (All Robinson quotes are from this talk.)

Chapter 16: What Language Does Your Body Talk?

Distances appear farther: M. Balla and D. R. Proffitt, "Visual–Motor Recalibration in Geographical Slant Perception," *Journal of Experimental Psychology Human Perception and Performance* 25, no. 4 (August 1999): 1076–1096, https://doi.org/10.1037//0096-1523.25.4.1076.

better putts: Jessica Witt et al., "Putting to a Bigger Hole: Golf Performance Relates to Perceived Size," *Psychonomic Bulletin and Review* 15, no. 3 (June 2008): 581–585 https://doi.org/10.3758/pbr.15.3.581; Jessica Witt, Sally Linkenauger, and Dennis Proffitt, "Get Me Out of This Slump! Visual Illusions Improve Sports Performance," *Psychological Science* 23, no. 4 (March 2012): https://doi.org/10.1177/0956797611428810.

change with your heartbeat: Saeedeh Sadeghi, Marc Wittmann, Eve DeRosa, and Adam K. Anderson, "Wrinkles in Subsecond Time Perception Are Synchronized to the Heart," *Psychophysiology* 60, no. 8 (August 2023): https://doi.org/10.1111/psyp.14270.

People shown a fearful face: Sarah Garfinkel et al., "Fear from the Heart: Sensitivity to Fear Stimuli Depends on Individual Heartbeats," *Journal of Neuroscience* 34, no. 19 (May 2014): 6573–6582, https://doi.org/10.1523/JNEUROSCI.3507-13.2014.

"cognitions and emotions": British Neuroscience Association, "How Our

Bodies Interact with Our Minds in Response to Fear and Other Emotions," *ScienceDaily*, April 2013, www.sciencedaily.com/releases/2013/04/130407211558.htm.

the process: Brian Hseueh et al., "Cardiogenic Control of Affective Behavioural State," *Nature* 615 (March 2023): 692–699, https://doi.org/10.1038/s41586-023-05748-8; The process used: "We first achieved cardiomyocyte-restricted expression by placing the ChRmine transgene under the control of the mouse cardiac troponin T promoter (mTNT), using the AAV9 serotype, which exhibits tropism for cardiac tissue. Infection of cultured primary cardiomyocytes with AAV9-mTNT::ChRmine-2A-oScarlet enabled light-evoked contractions with irradiance as low as 0.1 mW mm−2, consistent with the photosensitivity of ChRmine in neurons."

"stress or fear": Bethany Brookshire, "In Mice, Anxiety Isn't All in the Head. It Can Start in the Heart," *Science News*, March 2023, https://www.sciencenews.org/article/mice-anxiety-brain-heart-emotion.

"emotional or affective states": Hseueh, "Cardiogenic Control."

their abilities on a test: Barbara Fredrickson et al., "That Swimsuit Becomes You: Sex Differences in Self-Objectification, Restrained Eating, and Math Performance," *Journal of Personality and Social Psychology* 75, no. 1 (July 1998), 269–284, https://doi.org/10.1037//0022–3514.75.1.269.

associated with a doctor: Hajo Adam and Adam Galinsky, "Enclothed Cognition," *Journal of Experimental Social Psychology* 48, no. 4 (July 2012): 918–925, https://doi.org/10.1016/j.jesp.2012.02.008.

couldn't replicate the findings: Devin M. Burns et al., "An Old Task in New Clothes: A Preregistered Direct Replication Attempt of Enclothed Cognition Effects on Stroop Performance," *Journal of Experimental Social Psychology* 83 (July 2019): 150–156, https://doi.org/10.1016/j.jesp.2018.10.001.

the core principle: Hajo Adam, Adam Galinsky, "Reflections on Enclothed Cognition: Commentary on Burns et al.," *Journal of Experimental Social Psychology* 83 (July 2019): 157–159, https://doi.org/10.1016/j.jesp.2018.12.002.

Men who put on: Michael W. Kraus and Wendy Berry Mendes, "Sartorial Symbols of Social Class Elicit Class-consistent Behavioral and Physiological Responses: A Dyadic Approach," *Journal of Experimental Psychology* 193, no. 6 (Dec 2014): 2330-40, https://doi.org/10.1037/xge0000023.

"new ideas in the mind": Karen Pine, "Mind What You Wear Because It Could Change Your Life," *Sheconomics* blogspot (blog), May 2014, http://sheconomics.blogspot.com/2014/05/mind-what-you-wear-because-it-could.html.

"**If Interstate 101**": Alan Toth, "The Science Behind Muscle Memory," Stanford Medicine Scope, July 2022, https://scopeblog.stanford.edu/2022/07/15/the-science-behind-muscle-memory/.

cellular changes linger: K. Gunderson et al., "Muscle Memory: Virtues of Your Youth?", *Journal of Physiology* 596, no. 18 (September 2018): 4289–4290, https://doi.org/10.1113/JP276354.

"**Every avenue we tried**": Camonghne Felix, "Simone Biles Chose Herself," *Cut*, September 27, 2021, https://www.thecut.com/article/simone-biles-olympics-2021.html.

"**do not recommend**": Sophie Lewis, "Simone Biles Opens Up About Withdrawal from Olympic Competitions: 'I Don't Think You Realize How Dangerous This Is,'" CBS News, July 30, 2021, https://www.cbsnews.com/news/simone-biles-olympics-gymnastics-withdrawal-twisties/.

Chapter 17: The Body-Mind Happiness Plan

"**they ought to be seized**": Frédéric Gros, *A Philosophy of Walking*, trans. John Howe (Verso Books, ed. 2015), 100.

Index

A
Ackerman, Jonathan, 10
addictions, 4–5
ADHD, 254, 266, 300–302
adrenaline, 114, 116, 330
Aduriz, Adoni Luis, 126
aesthetics, 66–67
afferent neurons, 34
afterburn, 209
aliefs, 132
American College of Sports Medicine, 198–199
America's Next Top Model (TV show), 36–37
anandamide, 204–205
Angelou, Maya, 23
anger, 41–46
antibiotics, 263–264
antidepressants, 258–261, 265
anxiety-to-excitement technique, 52–54, 55
Apple Watch, 3
Archimedes, 282–283
architecture, 109–113
Aristotle, 71, 274–275
Arndt, Kenneth, 31, 37, 38
aromatherapy, 240
arousal, physiological, 47–50
arthritis, 226
artificial intelligence (AI), 26–27, 290–293, 295
autonomic nervous system, 28–29
The Awakening (Chopin), 96

B
Baker Baker paradox, 135
Ball, Lucille, 50
ball-tossing game, 16–17
Banks, Tyra, 36–37
Bargh, John, 6–7, 12–14, 18
Barrett, Lisa Feldman, xvi, 51
Barron, Carrie, 75
Basner, Mathias, 113–114, 115–116, 118
Beecher, Mark, 86
behavior, body-first approach to, xiv–xix
behavioral science, 21
Berman, Marc, 84–85
Bernardi, Luciano, 118
beta-blockers, 28–29, 59
The Big Bang Theory (TV show), 18

Index

Biles, Simone, 325–327
biofeedback loops, 21
biophilia hypothesis, 100–101
birdsong, 119
Bittman, Mark, 168, 170
blue spaces, 92–100, 335
Blumenthal, Heston, 127–128
bodies
  and behavior, xiv–xix
  brain-body connection, xviii–xix
  failures of, xiii
  feelings signaled by, xii–xiii, 314–316
  listening to, xix–xx
  motor memory, 322–327
  thinking process in, 298–303
  understanding signals of, 330–331
body neutrality, 142–143, 156
body positivity, 141–142
Botox, 29–33, 37–38
The Botox Diaries (Kaplan & Schnurnberger), 32
brain
  amygdala, 224
  brain-body connection, xviii–xix
  cerebral mystique, 282–285
  frontal lobe, 133
  hippocampus, 61
  insula, 15–16
  medial orbitofrontal cortex, 124
  medial prefrontal cortex, 289
  as a prediction machine, 128–129, 133–134
  thalamus, 134, 219, 221
Brewer, Judson, 4–5, 170–171

Bronzaft, Arline, 115
Brooks, Alison Wood, 52–53
brown noise, 116–117
brutalism, 110
built environments, 109–113
Buonomano, Dean, 327
Burke, Edmund, 33
Bushman, Brad, 41–45

C

Campbell, Gregory, 235
carpe diem, 146–150
catharsis theory, 41–46
cathedrals, 111–112
Ceaușescu, Nicolae, 152
central nervous system, xvii–xviii
Cézanne, Paul, 104–105, 109
Chandon, Pierre, 173–175, 178
ChatGPT, 291
Chopin, Kate, 96
Chou, Roger, 217
Churchill, Winston, 103
Cinnabon, 63
Cirino, Nicole, 160, 164
Clark, Andy, 299
clothing, 317–322, 333
clutter, 107
Cole, Kat, 63
Cole, Nat King, 36, 38
Coles, Nicholas, 35
comfort foods, 189–192
confirmation bias, 262–265
consciousness
  bodily, xii–xiii, xvi
  brain, xiii, xvii–xviii
  speed of awareness, 11–12
convergent thinking, 280

Index

Corden, James, 65
cortisol, 85, 93, 114, 116, 330
coziness, 332–333
creativity
  body and, 298–303
  cerebral mystique and, 282–285
  movement and, 271–280, 305–307
cross modalities, 123
Crum, Alia, 180–181
Cumberbatch, Benedict, 76
curiosity, 4–6
current self, 143–150

D

Damasio, Antonio, xv, xvi, xvii, 46
dancing, 278–280, 304–307
Dante, 15
Darwin, Charles, 32, 33, 44
Dead Poets Society (film), 145
DeBrusk, Jake, 235
Deisseroth, Karl, 314–315
De Kooning, Willem, 278
depression treatments
  antidepressants, 258–261, 265
  exercise, 206–209
  placebo effect, 258–261
  posture and, 333–334
Descartes, René, xvi, 67, 143
Diderot, Denis, 66
Didion, Joan, 299
dietary supplements, 250–252
Dietrich, Arne, 204
Ding, Jun, 323
distraction, 238–242, 303–304
divergent thinking, 272–273, 275–276, 280

Dockterman, Eliana, 163
Don't Forget the Bacon (Hutchins), 136–137
doodling, 288–290
dopamine, 73, 90, 159, 161, 193, 200, 204, 208, 281, 317
Dove Real Beauty campaign, 142
Downey, Robert, Jr., 76
Duchenne de Boulogne, Guillaume-Benjamin-Amand, 37
Duchenne smile, 37
Dweck, Carol, 57

E

Eagleman, David, 12
earthworms, xvii
eating
  body awareness and, 167–169, 186–189
  comfort foods, 189–192
  diets, 170, 179–180, 182–184
  hunger hormone, 180–181
  instinctive, 184
  pleasures of, 73, 144, 175–181, 338–339
  satiety, 169–171
  sensory aesthetics and, 125–128
  sensory-specific satiety, 172–175
Edison, Thomas, 301–302, 307
embodied cognition
  temperature effects on, 13–19
  weight effects on, 6–13
emotions. *See* feelings and emotions
enclothed cognition, 319
endocannabinoids, 160–161

endorphins, 73, 161, 162, 200, 203–205, 337
energy preservation instinct, 196–200
ephemeral phenomena, 101–102
Epicurus, 173
Etkin, Jordan, 2–3
Euclid, 274
"Eureka!" moments, 282–284
Evans, Chris, 163
evolutionary biology, xiii
exercise
    creative inspiration and, 272–280, 305–307
    energy preservation and, 196–199
    hormones, 203–205
    learning to love, 200–203
    metabolism and, 209–211
    mood and, 144, 205–209
    pleasures of, 337–338
    research on, 202–203

F

facial expressions
    emotions and, 32–34
    frowning, 30–33, 37–38
    smiling, 34–38, 333–334
facial feedback hypothesis, 33–34, 35
fear, 313–316
feelings and emotions
    arousal congruent, 53–54
    bodily signals of, xii–xiii, xvii–xviii, 314–316
    curiosity about, 4–6
    facial expressions and, 32–33, 333–334

    misattribution of, 46–48, 88, 122
    and pain, 221–222
    physical sensations of, 70–72
    reframing, 54–59, 336–337
    repression of, 44–45
    sources of, xv
    two-factor theory of, 49
fibromyalgia, 226, 247
fidgeting, 303–304
fingertips, 154–156, 292
Finland, 106
Fitbit, 1–3
fixed mindset, 57–58
flowers, 90–92
Foreign Affairs (Lurie), 164–165
The Fractal Geometry of Nature (Mandelbrot), 108
fractals, 107–109, 112
Freud, Sigmund, 40–41
frowning, 30–33, 37–38
future self, 143–147

G

Gaiman, Neil, 75–76
Garfinkel, Sarah, 313–314
gate-control theory, 248–249
Gaudí, Antoni, 112
Gendler, Tamar Szabo, 132
ghrelin, 180–181
Gilbert, Dan, 147
Gladwell, Malcolm, 273
Google, 2
Grand Canyon Skywalk, 131–132
gratitude, xi, xx, 19, 25, 30, 70, 84, 90–91, 104, 111, 152–153, 203, 328–329, 341

Index

Gros, Frédéric, 341
growth mindset, 57–58

H

Handel, Barbara, 277
Happy Body Lifestyle, 184–185, 188–189, 339{~?~EP: Query with author about "Healthy Body Lifestyle"}
haptic mindset, 10–13
Harvey, William, 71
health trackers, 1–3
heart, as center of emotion, 71–72
heartbeat, 313–316
Hennenlotter, Andreas, 37–38
Herder, Johann Gottfried von, 66, 152
Herz, Rachel, 61
The Hidden Life of Trees (Wohlleben), 89
hippocampus, 61
Hippocrates, 65, 255
hobbies, 73–75
Horace, 147–148
hormones, xiii–xiv, 73–75, 159–162, 180–181, 203–205
Hutchins, Pat, 136
hygge, 332

I

IJzerman, Hans, 16–17
Inferno (Dante), 15
insula area of brain, 15–16
intelligence
 artificial, 26–27, 290–293, 295
 human, 292–296
 morphological, 295–296
 physical, 296–298

International Style, 110
interoception, xv–xvi
Isbister, Katherine, 304
Ishiguro, Kazuo, 26

J

Jack Daniels, 125
James, Henry, xvi
James, William, xvi–xvii, 314, 316
Jamieson, Jeremy, 55–59
Jasanoff, Alan, 281–282
Jefferson, Thomas, 45
jellyfish, 67–70
Johnson, Virginia, 159–160, 165
Jones, Jeannete Haviland, 91

K

Kaimal, Girija, 289
Kant, Immanuel, 340
Kaptchuk, Ted, 254–255
Karlesky, Michael, 304
King, Stephen, 299
Klara and the Sun (Ishiguro), 26
Kline, Franz, 278
knitting, 73
Kreuther, Gabriel, 126
Krznaric, Roman, 147–148, 150

L

Lakoff, George, 9, 23
Lavidis, Nick, 62
Le Corbusier, 110
Legos, 74–75, 294
Li, Fei-Fei, 295–296, 307
Lieberman, Daniel, 195, 196–199
limbic system, 60–61

Index

Lurie, Alison, 164
Lynne, Gillian, 306–307

M

MacKerron, George, 82–84
Mackey, Sean, 220–221, 223, 228, 230–232
Major League Eating, 169
Mandelbrot, Benoit, 108
Mappiness project, 83–84
Marley, Bob, 240
Marvell, Andrew, 150–151
The Marvelous Mrs. Maisel (TV show), 20
massage, 64–66
Masters, William, 159–160, 165
Masters of Sex (TV show), 159
medical care, 236–265
memory
   context-dependent, 286–287
   motor, 322–327
   sensory cues and, 134–137
   smell and, 60–62
   taste and, 190–191
menu descriptions, 177–178
Mestral, George de, 284
metabolism, 182, 209–210
Metaphors We Live By (Lakoff), 9
Michel, Charles, 125
Microsoft, 291
Miller, Lewis, 90
mind-body duality, xvi, xix–xx
mindset, fixed vs. growth, 57–58
mirror neurons, 33
misattribution, 46–48, 88, 122
Mitchell, Manteo, 234–235
Mochon, Daniel, 74

Monet, Claude, 104, 105, 109
morphology, 295–296
motor memory, 322–327
Mulaney, John, xii
multivitamins, 250–253
Murthy, Venkatesh, 60
muscle memory, 322–327
muscular tension, 24–25
Myers-Briggs test, 286–287

N

Nathan's Famous International Hot Dog Eating Contest, 169
National Institutes of Health, 246–247
natural environments, 82–85, 334–335
Needham, Dale, 158
neurons, 6–8, 24, 33–34, 51, 58, 105, 155–158, 219, 222, 224, 226, 296–297, 332–333
neurotransmitters, 65, 88, 89, 161, 162, 191, 200, 202, 203, 207, 249, 255, 257, 330, 337
Newton, Isaac, 283–284
Nietzsche, Friedrich, 340–341
noise, 113–119
norepinephrine, 159, 208
Norton, Michael, 74

O

obesity, 182, 194
observation decks, 129–132
O'Connor, Flannery, 299
OpenAI, 291
opioids, 161, 246–247
Oppezzo, Marily, 271–273, 280

optogenetics, 314–316, 317
orgasm, 159–162
orientation, physical, 9–10, 23–24
outdoors, spending time, 82–85, 334–335
oxygen, 82
oxytocin, 64–67, 73, 89, 159, 161, 162

P

pain
    back, 215–219, 229–230, 231–232, 240
    catastrophizing, 242–244
    central sensitization, 226–228
    chronic, 225–228
    constructed by the brain, 228–230
    distraction and, 238–242
    effects of worry on, 237–238
    ignored by brain, 234–237
    magic pill strategy, 244–246
    neuropathic, 227
    nociceptive, 227, 247
    physical, 215
    placebo effect and, 256–261
    reality of, 223–225
    subjective experience of, 219–223
paleo diet, 183–184
Palladino, Lucy Jo, 301–302
Park, James, 1–2
pedometers, 1–3
Peper, Erik, 21–25
perception, 311–313
perspective changing, xii
Pert, Candace, 203–204
Pfeifer, Rolf, 291–294, 296
physical system, 292–295
Pine, Karen, 320–321
pink noise, 116–117
place, sense of, 103–106
placebo effect, 35, 38, 42, 181, 253–261, 265–267
Plato, 26–27, 274–275
Poe, Edgar Allan, 32–34
Pollock, Jackson, 277–278
positivity, 328–329
posture, 20–25
Proust, Marcel, 190–191
psychophysiology, 21
Ptolemy, 274
"The Purloined Letter" (Poe), 32–33
Pythagoras, 274

R

Rage Rooms, 39–40
rainbows, 101–102
Raphael, 274–275
real you, 75–77, 287
reappraisal process, 55–56
reframing, xii, 54–59, 336–337
Reichl, Ruth, 187
Remembrance of Things Past (Proust), 190–191
Robinson, Ken, 304–307
robotics, 292–295
roller coasters, 48–50
Rosenberg, Harold, 278
Rossetti, Christina, 71
Rousseau, Jean Jacques, 340
Rozin, Paul, 132
Rumi, 103
rumination, 44, 237–238, 242–243
runner's high, 204

S

Sagrada Familia, 111–112
Sandini, Giulio, 292
satiety, 169–171, 172–175
scenicness, 105
Schachter, Stanley, 49
Schnurnberger, Lynn, 32
The School of Athens (Raphael), 274–275
screen time, 148–149
Scully, Vincent, 109
Seagram Building, 109–110
Seasonal Affective Disorder (SAD), 86
Segovia, Andreas, 60
seizing the day, 146–150
self-efficacy, 208–209
senses and sensory input
    aesthetics and, 66–67
    awareness of, 81–82
    as distraction from pain, 239–240
    environmental input and, 76–77
    internal, 70–72
    learning and, 298–303
    location and, 122–128
    present moment experiences of, 144–145
    sight, 129–134
    smell, 60–64, 240
    sound, 113–119, 127–128
    taste, 122–128, 190–191, 224–225
    touch, 64–67, 150–156
sensory neurons, 154–158, 219, 222
sensory-specific satiety, 172–175
serotonin, 73, 86, 88, 159, 161, 208, 260
sex, 73, 142–143, 158–166
Shakespeare, William, 71
Sherlock Holmes, 76
Simard, Suzanne, 89
Sin City Smash, 39–40
Singer, Jerome, 49
Sitti, Metin, 296
sitting, 194–195, 277
smell
    and memory, 61
    effect on mood and emotion, 60
    power of, 63–64
    to treat pain, 240
smiling, 34–38, 333–334
smizing, 36–37
Socrates, 26, 274
soft fascination, 84, 93
somatosensory network, 72
sounds, 113–119, 127–128
Spence, Charles, 123–127
Spencer, Percy, 284
Strack, Fritz, 34–35
stress hormones, 85, 93, 114, 116–117
stress response, 56–57, 239, 330–331
stress/threat analgesia, 236
stretching, 334
sunsets, 101–102
sunshine, 85–88
supplements, 250–253
survival instinct, 67–70
Sweeney, Julia, 176

T

taste, sense of, 122–128, 190–191, 224–225
Taylor, Richard, 108, 110
temperature, effects of, 13–19
THC, 204
therapy dogs, 157–158
Thoreau, Henry David, 89, 272
"To His Coy Mistress, To Make Much of Time" (Marvell), 150–151
touch
  physical, 150–154
  pleasures of, 156–158
  sense of, 64–67, 154–156
treadmills, 195–196
tree hugging, 89–90
two-factor theory of emotion, 49
ultra-processed foods, 185–186
unconscious bias, 7

V

vacations, 81–82
vagus nerve, 72
venting theory, 41–46
Venturi, Robert, 110–111
visual inputs, 129–134

W

Wager, Tor, 229, 232, 236, 256–258
walking, 1–3, 199, 273–278, 340–341
warmth, effects of, 13–19
water, 92–100, 335
weather, 85–88
Weber, Andrew Lloyd, 306
weight, effects of, 6–13
weight loss, 182–183, 209–211
Wham It!, 43
whiskey, 124–125
White, Mathew, 94–95, 97–100, 101–102
white noise, 116–117
Wilde, Oscar, 1
Williams, Robin, 145, 146
Wilson, E.O., 100
Wilson, Rainn, 98
wine, 122–124
Winterson, Jeannette, 25–26
Witt, Jessica, 313
Wohlleben, Peter, 89
Woolf, Clifford, 224–225
Wordsworth, William, 340
World Happiness Report, 105–106

Z

Zuckerberg, Mark, 320
Zuker, Charles, 224–225
Zwick, Edward, 163–164

About the Author

Janice Kaplan is a journalist, TV producer, and the author of fifteen popular books including the New York Times bestseller *The Gratitude Diaries* and *The Genius of Women*. Janice was editor-in-chief of *Parade* magazine and the creator and executive producer of more than thirty primetime network television specials. She has appeared regularly on national TV shows including *Today* and *Good Morning America* and is a frequent guest on podcasts and radio shows. An energetic speaker at events around the country, she graduated from Yale University and lives in New York City.